REGENERATING ROMANTICISM

Regenerating Romanticism

Botany, Sensibility, and Originality in British Literature, 1750–1830

Melissa Bailes

University of Virginia Press
Charlottesville and London

University of Virginia Press
© 2023 by the Rector and Visitors of the University of Virginia
All rights reserved
Printed in the United States of America on acid-free paper

First published 2023

1 3 5 7 9 8 6 4 2

Library of Congress Cataloging-in-Publication Data
Names: Bailes, Melissa, author.
Title: Regenerating Romanticism : botany, sensibility, and originality in British literature, 1750–1830 / Melissa Bailes.
Description: Charlottesville : University of Virginia Press, 2023. | Includes bibliographical references and index.
Identifiers: LCCN 2022035690 (print) | LCCN 2022035691 (ebook) | ISBN 9780813949406 (hardcover ; acid-free paper) | ISBN 9780813949413 (paperback ; acid-free paper) | ISBN 9780813949420 (ebook)
Subjects: LCSH: Science in literature. | English literature—18th century—History and criticism. | English literature—19th century—History and criticism. | Sensitivity (Personality trait) in literature. | Botany in literature. | Romanticism—Great Britain. | Literature and science—Great Britain—History—18th century. | Literature and science—Great Britain—History—19th century. | Scientific literature—Great Britain—History—18th century. | Scientific literature—Great Britain—History—19th century. | LCGFT: Literary criticism.
Classification: LCC PR448.S32 B35 2023 (print) | LCC PR448.S32 (ebook) | DDC 820.9/007—dc23/eng/20230113
LC record available at https://lccn.loc.gov/2022035690
LC ebook record available at https://lccn.loc.gov/2022035691

Cover art: "Night-Blowing Cereus, or Cactus Grandiflorus," Robert John Thornton, *New Illustration of the Sexual System of Carolus von Linnaeus*, 1807 (Image courtesy of the Linda Hall Library of Science, Engineering & Technology); *Mad Kate*, Henry Fuseli, 1806/7 (Image courtesy of Frankfurter Goethe-Museum, © Ursula Edelmann—ARTOTHEK); *Pancratium maritimum*, Mary Delany, *Flora Delanica*, 1775 (© The Trustees of the British Museum)

CONTENTS

Acknowledgments · vii

Introduction: Revealing the Straw Man, or
The Historical Hoodwinking of Romanticism · 1

Part I. Temporal Sensibilities: Circannual and Circadian Rhythms

1 Botany's Seasonal Disorder: Thomson's Progressive Time, Conjectural Histories, and the Backwardness of Spring · 35

2 Linnaeus's Botanical Clocks: Chronobiological Mechanisms in the Scientific Poetry of Erasmus Darwin, Charlotte Smith, and Felicia Hemans · 61

Part II. Sensibility and Empire: Gender, Race, and Nation

3 Transformations of Gender, Race, and Poetic Sensibility: Maria Riddell's Transatlantic Botany and Biopolitics · 87

4 Cultivated for Consumption: Botany, Colonial Cannibalism, and National/Natural History in Sydney Owenson's *Wild Irish Girl* · 110

Part III. In/effability: Sensibilities of Description, Classification, and Defiance

5 "On the Green Margin": Place, Sensibility, and Originality in Charlotte Smith's "Flora" · 133

6 Botany and Madness: Anna Seward, Sensibility, and the Floral Insanities of Darwin, Cowper, Wordsworth, and Clare · 155

Conclusion: De Quincey, Hazlitt, Wordsworth, and the
Critical Fate of Romanticism and Scientific Literature   187

Notes   209

Bibliography   241

Index   263

ACKNOWLEDGMENTS

Several fellowships made possible the research and writing of this book. I am especially grateful for a Ruth W. and A. Morris Williams Jr. Fellowship from the National Humanities Center, granting a year of much-needed time and resources to bring this study to completion. Both the Linda Hall Library and the Lewis Walpole Library provided short-term fellowships, allowing access to rare sources that enrich these pages. Thanks also to Tulane University and the Dean of the School of Liberal Arts for a book subvention. Portions of chapters 3 and 5 were published as articles in *Eighteenth-Century Fiction* (2019) and in the edited collection *Placing Charlotte Smith* (Lehigh University Press, 2020), as were early versions of chapters 2 and 4, respectively, in *Studies in Romanticism* (2017) and *Eighteenth Century: Theory and Interpretation* (2018). Many thanks to Alan Bewell, Geraldine Friedman, and Noah Heringman for their letters of recommendation and notes of encouragement as I pursued fellowship support. Geraldine also provided very helpful feedback on two chapters. Jennifer Keith first got me thinking about women writers and natural history during my early years of graduate work; she generously read chapters of this book, and I have owed more than I can say to her friendship and kindness over the years. Allison Davis has been a wonderful and constant, though geographically distant, companion throughout the writing of this book: reading almost the entire manuscript, sharing life events, and making everything more fun. Meghan Costello-Ishak, Jessica Damián, Kathryn Hansen, Annie Hayes, Laura Merchant, Donnell Oakley, Kristen Pond, Alice Touchette, and Anna White enlivened my work through meaningful conversations, care, and community.

My amazing editor, Angie Hogan, again believed in my work, and I truly appreciate her help and excitement in bringing this project to fruition. Thanks also to my two anonymous readers, and everyone at UVA Press who contributed to the completed book. I am extremely grateful to the British Society for Literature and Science, as well as the British Association for Romantic Studies, for endorsing my earlier book, *Questioning Nature*, on which *Regenerating Romanticism* builds. Ted Underwood, Robert Markley, Justine Murison, and Gillen Wood read a nascent version of chapter 5, on Charlotte Smith's "Flora," many years ago, and I remain thankful for their excellent feedback and support for my ideas; Ted, with his characteristic graciousness and insightful suggestions, read two additional chapters of this book. In 2014, Carl Thompson generously invited me to present part of chapter 3 in a colloquium on women's travel writing at the Senate House Library in London, where I was fortunate to meet and feel inspired by several of the titans in botanical studies. I am grateful to Joshua King and Kristen Pond for inviting me to deliver an early version of chapter 1 at the "Religion and Ecology in the Nineteenth Century" conference at Baylor University in 2019, and to Susan Oliver and other fellow presenters who offered helpful questions and feedback. I also wish to thank all of my colleagues in the English department at Tulane University, as well as Kate Adams, Tom Albrecht, Tita Chico, Mike Kuczynski, Adam McKeown, Julie Park, Ruth Perry, Jane Pinzino, Molly Rothenberg, Barb Ryan, Katie Sagal, John Sitter, Courtney Weiss Smith, Danielle Spratt, Ed White, and Eugenia Zuroski for their words and gestures of help or support, large and small, along the way. Finally, much gratitude goes to my family, Sue and Eric Bailes, Katie, Ethan, and Emma Kleisch, and Sean Orgias, for their love and patience, and for enthusiastically joining in some of my botanical explorations as I wrote this book.

REGENERATING ROMANTICISM

Introduction

Revealing the Straw Man, or The Historical Hoodwinking of Romanticism

REGENERATING ROMANTICISM argues that this field of literary studies is currently built on a number of false premises relating to science and scientific literature. As I will show, William Wordsworth and various other male writers during this era often sought to distinguish what would become canonical Romantic-era literature from the late eighteenth- and early nineteenth-century movement of scientific literature by targeting the poet, physician, and naturalist Erasmus Darwin. In fact, in an 1802 letter, Samuel Taylor Coleridge indicated that differentiating Wordsworth's poetic theories and methods from Darwin's had been the goal of the preface to *Lyrical Ballads*, declaring that "Darwin and Wordsworth" had each given "a defense of *their* mode of Poetry" and asserting opposition between the two.[1] In a 1799 review of *Lyrical Ballads*, the *British Critic* confessed, "We infinitely prefer the simplicity, even of the most unadorned tale in this volume, to all the meretricious frippery of the *Darwinian* taste."[2] In 1818 the *Edinburgh Magazine* observed that, "in matter, and in manner, the Lake and Darwinian schools of poetry are the very antipodes of each other—hostile in all their doctrines, and opposite in every characteristic."[3] Such statements chart how individual authors of scientific literature became critically categorized and dismissed as belonging to the "Darwinian school," thereby eliding the fact that many of the imaginative authors, especially women, most obviously engaging with technical science also critiqued or avoided the same perceived shortcomings in Darwin's verse. Thus, within certain key texts discussing aesthetics during the Romantic era, Darwin became a "straw man" in the sense that aspects of his works were refuted as if his arguments or methods represented the larger movement of scientific literature, even when they did not. In this way, Wordsworth and some other now-canonical male writers and theorists denigrated the period's scientific literature as lacking the very qualities that it in fact sought to achieve, particularly originality and sensibility.

The success of this tactic in establishing what would become viewed as many of the canonical or "high" literary principles of Romanticism contributed to the historical devaluation of Darwin's works as well as those of numerous other male and, especially, female authors of scientific poems and novels; it also created a paradox of misunderstandings regarding the aesthetic intentions of the era's scientific literature that continues to hinder or mislead scholars of Romanticism even today. Indeed, I argue that, by making Darwin the straw man, antagonistic nineteenth-century critics, writers, and reviewers distracted from the fact that in many ways women writers—and especially the larger movement of scientific literature (which was strongly associated with women)—were the actual, principal targets of these attacks. Nevertheless, Darwin provided an easy, prominent mark for falsely grouping and appearing to refute what was, in reality, a diverse and multifaceted literary movement, comprised largely of women. As I will demonstrate, such strategies enabled some of the era's literary critics, particularly Thomas De Quincey, to portray literature and science as being in competition with one another while also establishing standards for the literary canon that in fact *mirrored* developing ideas of scientific or biological sexism and racism. These nineteenth-century critical approaches thereby sought to exclude from the canon not only this movement of scientific literature but also, relatedly, the works of women and non-Europeans more generally, with consequences that shaped and arguably continue to influence literary studies.

I began to correct these misperceptions regarding the originality of scientific literature in my previous book, *Questioning Nature: British Women's Scientific Writing and Literary Originality, 1750–1830*, and will address pertinent aspects of its argument in sections below. In *Regenerating Romanticism*, I shift my focus primarily to analyzing how sensibility relates to this period's movement of scientific literature. By "scientific literature," I refer to genres including poetry, novels, travel writing, children's literature, and even literary criticism written during the late eighteenth and early nineteenth centuries that obviously and technically engage with the natural sciences, and (for the purposes of this study) especially with botany. Also, in specifically calling this writing "scientific literature" and thus reinforcing its status *as* literature, I am taking a conscious stand against De Quincey's 1823 exclusionary definition of literature ("All that is literature seeks to communicate power; all that is not literature, to communicate knowledge"), which I analyze further in the conclusion, particularly in his formulation's bias against women's writing.[4]

In recent decades, there has been growing interest in the interactions between British literature and science during the Romantic period, and with good reason. Traditional narratives conceived of Romantic-era literature as growing out of an opposition to Enlightenment science and rationality. Thus, many scholars have sought to reveal ways in which late eighteenth- and early nineteenth-century literature in fact incorporated the sciences during this period when "science" altered from its broader connotations, which also included "philosophy" and "knowledge," to its modern, specialized meaning of, as Wordsworth put it in 1802, "what is now called science"; it is the latter, modern meaning that constitutes my usage of the term in this study.[5] Numerous scholarly monographs and collections trace how the now-canonical male authors of this period employed the sciences in their poetry and prose, and, more slowly, increasing attention has been paid to male and female authors who, subsequent to the era, became lesser known, but often engaged more obviously and directly with scientific fields. Indeed, although valuable in themselves, these attempts to locate canonical male authors' literary uses of the sciences often serve to reinforce those writers' dominance while excusing the parts that some of them played in the historical erasure of other authors, many of whom were women, whose literature focused more directly and prominently on the sociopolitical authority and aesthetic possibilities available through the natural sciences.

As scholars such as Marlon Ross and Anne Mellor began to demonstrate decades ago, the canonization of certain works of Romantic-era aesthetics historically established masculine and often misogynistic literary standards, specifically phrased to exclude women from recognition or participation in, for example, "the real language of men," or from the role of the poet, defined by Wordsworth as "a man speaking to men" (608, 603).[6] New historicists, such as Jerome McGann, urged scholars toward a healthy skepticism about authors' presentations of themselves and the events of their eras, especially when those writers may have had prejudices or a stake in the outcome, and the canon has since received further challenges from historical-materialist, Foucauldian, feminist, and postcolonial criticism, among other analytics.[7] However, despite late twentieth- and early twenty-first-century successes in opening the literary canon, so that, for instance, more women writers of the Romantic era are being taught than ever before, and historically less- or noncanonical male writers now receive greater attention, Romanticism arguably continues to uphold the works of its traditionally canonical "big six" male writers as its dominant or representative literary expression.[8] Likewise, there often remains a stigma of secondary status for the era's scientific literature, and for writings more generally, by its historically less- or noncanonical authors, especially women.

During the Romantic period, certain texts by writers including Wordsworth, Coleridge, Hazlitt, and De Quincey sought to distinguish literature from science primarily as a way of shoring up literature's cultural influence at a time when scientific fields became increasingly professionalized and recognized as a locus of cultural and sociopolitical authority. In the process, these works of what became canonical Romantic-era aesthetics also denigrated the period's literature (especially poetry) that most prominently and technically engaged with the sciences, indicating that such scientific literature lacks both originality and sensibility.[9] As I have stated, the most obvious target of these attacks was Erasmus Darwin, the grandfather of Charles. Darwin's long scientific poem, *The Botanic Garden*, consisting of two parts, *The Economy of Vegetation* (1791) and *The Loves of the Plants* (1789), in which the second was published first and versifies Carl Linnaeus's system of botanical classification, became "the most popular and the most controversial nature poem of the 1790s."[10] Although this poem initially gained great praise, its success increasingly drew vitriol from critics and reviewers as well as from rising poets seeking to distinguish their own work from Darwin's. Thus, while Wordsworth's preface to the *Lyrical Ballads* (1802), often now viewed as a de facto manifesto of the Romantic movement, critiques rival literary and scientific groups and individuals including the Della Cruscans, Gothic literature, and Humphry Davy, it takes aim at Darwin when denouncing perceived deficiencies in modern poetry and differentiating the Poet from the "Man of science," thereby distinguishing literary and scientific vocations.[11] Wordsworth's assertions regarding science subsequently became standardized through repetition in core texts used to establish canonical understandings of the period's aesthetics. This culminated in De Quincey's 1848 championing of the "literature of power," which he associates with sensibility and unindebted originality in poetry by Wordsworth, Milton, and Shakespeare, as superior to the didactic and often scientific "literature of knowledge," which he presents as lacking originality and feeling.

In the latter half of the eighteenth century, writers often emphasized concepts of originality and sensibility as reactions against modes of imitation and satire that had been privileged within literary movements during the Restoration and early eighteenth century. Indeed, sensibility and originality became so prevalent as literary aspirations throughout the decades leading up to the nineteenth century that, as I will discuss, when planning the *Lyrical Ballads* (1798), Wordsworth and Coleridge pronounced novelty and sympathy to be "the two cardinal points of poetry."[12] I first addressed originality within *Questioning Nature* because it is perhaps a less intuitive (or more paradoxical) goal of the era's movement of scientific literature, while any scholar who has read the poetry of, say, Charlotte Smith, must recognize that scientific literature often blended with sensibility in this period. Nevertheless, it is crucial to explore sensibility because this constitutes the main quality, in addition to originality, that Wordsworth and some contemporary poets, critics, and reviewers found lacking in Darwin's scientific poetry, and that many writers of scientific literature centrally employed. In fact, writers such as James Thomson, William Whitehead (poet laureate, 1757–85), John Scott, Anna Seward, Smith, Anna Barbauld, George Crabbe, William Cowper, and Helen Maria Williams, among others, published poems of sensibility that engaged with natural history (comprising the fields of botany, zoology, and geology) before Darwin published *The Loves of the Plants*, and each of the women listed subsequently critiqued the dearth of feeling in his poetry.[13] Therefore, the lack of sensibility that Wordsworth sought to assign to Darwin's scientific poems cannot be applied to scientific literature more generally, and yet this is exactly what occurred over the course of the Romantic period.

In the late eighteenth and early nineteenth centuries, knowledge of natural history and expressions of sensibility each separately could serve as a means to literary and cultural authority and were often combined in texts by both male and female authors. This ability to meld scientific knowledge and the moral appeal of sensibility was particularly important for women writers at a time when women did not possess institutional political power. As I showed in *Questioning Nature*, many writers, including women, also blended literature and natural history in order to achieve literary originality. However, crucial to my argument in *Regenerating Romanticism* is that some now-canonical male writers of the era, such as Wordsworth and De Quincey, align both sensibility and originality with Wordsworth's poetic methods and, at the same time, detach each of these qualities from science, effectively severing writers of scientific literature, especially women, from their sources of literary and cultural authority. In other words, certain texts that retroactively became viewed as representative of "Romantic-era aesthetics" provided the ideologies for excluding from the canon many male and female writers of scientific literature by inaccurately denying such works' capacity for either originality or sensibility; those now-canonical texts additionally precluded many women writers more generally by appropriating sensibility and originality for masculinized theories of poetics and rejecting women's and (in the case of De Quincey, for instance) non-Europeans' ability to achieve either of these qualities in their supposedly "ideal" manifestations.

In this book, I argue that the key failings that some nineteenth-century critics, reviewers, and now-canonical male authors attributed to Darwin's poetic works need to be reevaluated and can no longer serve as an excuse to ignore or discount the era's writers of scientific literature, many of whom were practicing different aesthetics. Therefore, I am proposing the need to decenter—or at least deemphasize—conceptions of Darwin and his oeuvre as they have been stereotyped and applied to scientific literature of the period more generally. This is not to deny the importance of Darwin's works and his very real influence on the era's subsequent writers. In fact, Darwin appears, if only briefly, in every chapter of this book, and, as I demonstrate, cases can certainly be made for both his originality and his participation in various registers of sensibility. However, after his poetic publications, scientific literature was often confined either overtly or through implication under the broad category of "the Darwinian school," especially by some literary critics and reviewers during the first half of the nineteenth century; those critics and reviewers thereby erased authorial differences in order to consign such scientific texts either to oblivion or to the status of "secondary literature" in ways that continue even now to reinforce erroneous narratives about whose work counts as achieving originality and sensibility within the era's aesthetics. Thus, I contend that each author instead must be understood and appreciated individually and on his or her own terms.[14]

Additionally, I am arguing for the need to reevaluate or critique the aesthetic theories of nineteenth-century literary critics, reviewers, and now-canonical male writers, including Wordsworth, Coleridge, Hazlitt and, particularly, De Quincey, when these theories and works misrepresent (either overtly or by implication and effect) women, race, and/or science. These male Romantic-era writers' historical influence in shaping the literary canon has long been recognized; and, while some scholars have already gestured toward ways in which certain texts by these authors contributed to establishing or justifying canonical exclusions based on factors such as sex, race, and class, the consequences of such works for and in connection with the era's movement of scientific literature have yet to be fully acknowledged or explored. My study aims to shed new light on this complicated history and to call for greater accountability for these major male Romantic-period writers' historical appropriations, manipulations, and erasures of immediately preceding and concurrent literary movements, and to do so by examining the significance of scientific literature's relationship with modes of sensibility. Thus, this book's chapters primarily analyze the ways in which individual male, and especially female, writers variously interconnected botany and sensibility within their scientific literature. However, it is necessary first to provide additional historical context that more thoroughly explains the importance of originality, sensibility, and botany within the era, as well as how and why Darwin became the straw man within particular works of Romantic-era aesthetics, therefore further displaying his function as both an inspiration and a liability for many other authors of scientific literature.

Laying the Groundwork: Originality and Sensibility in Aikin and Darwin

Since *Regenerating Romanticism* builds on certain arguments made in my book *Questioning Nature*, it is useful to briefly summarize some of those assertions here. In the mid-eighteenth century, British literary critics and authors often complained of modern poets plagiarizing from classical writers as well as from one another, and thus of a seeming scarcity of new subjects for verse. However, as scientific discoveries and advancements particularly began to thrive in natural history at this time, some literary critics and theorists encouraged the melding of literature and science as a means to poetic originality. John Aikin, the physician and literary critic as well as brother of Anna Barbauld, published perhaps the best-known and most influential espousal of this critical exhortation in his *Essay on the Application of Natural History to Poetry* (1777). In these decades leading up to the nineteenth century, naturalists racing to describe and classify biological organisms called for the assistance of amateur observers whose knowledge of local ecologies could help map the natural order. Echoing this call to nonspecialists, Aikin encouraged poets to further scientific endeavors through study and close observation of natural phenomena both to assist naturalists and to reveal original subjects for verse.[15] According to Aikin, poets thus could avoid plagiarism and imitation, as well as inaccuracies in depicting the natural world, by displaying "original genius" through "attentive observation [of natural objects], conducted upon somewhat of a scientific plan."[16] For him, the poet-naturalist's observations thereby achieve the quality of lasting truth that is accorded to scientific principles, for "nothing can be really beautiful which has not truth for its basis."[17]

As I have discussed, the English georgic's seeming disappearance after 1767 functions as a dissemination that may be traced through the subsequent era's scientific literature.[18] James Thomson's popular georgic *The Seasons* (1726–46) proved especially influential in this regard, and Aikin praises Thomson as "the Naturalist's Poet" for his observations of the natural world that exemplify how poets may become better, more knowledgeable naturalists with respect to certain subjects than the naturalists themselves.[19] While acknowledging Thomson's influence, writers such as Maria Riddell, Anna Seward, and Charlotte Smith also claimed greater scientific and technical knowledge and authority than the older poet so that they positioned themselves as the inheritors of this scientific strain of literature, negotiating its dual emphasis on pleasure and truth, which sometimes manifests didactically. Indeed, many male and female writers participated in this blending of literature and natural history in pursuit of originality, not only in poetry but also in novels, children's literature, travel literature, literary criticism, and periodicals. These imaginative authors often contextualized their scientific theories and observations in relation to those of established male naturalists, paradoxically realizing originality through intertextuality in forms such as citation and borrowing, footnotes and endnotes, and reference and translation. Scientific texts by naturalists themselves generally represented collective, collaborative undertakings, building on, referencing, and correcting other naturalists' systems, theories, and observations. The writers of imaginative, scientific literature, too, often question, expand, or confirm naturalists' claims in their quests to achieve originality. Moreover, women's participation in the natural sciences established such strong associations between women and natural history in the public imagination that not only did Wordsworth and other male Romantics actively seek to separate their verse from this movement of scientific literature but also, by the 1820s and 1830s, male scientific professionals diligently labored to reinstate the masculinity of studying the natural sciences.[20]

Over the course of the late eighteenth and early nineteenth centuries, certain now-canonical male writers sometimes presented their texts as unindebted, and critical favor shifted more pronouncedly toward notions of autonomous authorship. Especially during the 1820s and 1830s, some literary critics and reviewers retroactively simplified and mythified the aesthetic fantasy of the solitary genius, selectively overlooking the intertextuality of works by Wordsworth and other male Romantic-era writers to instead emphasize these texts as constituting a new kind of poetry. In this way, certain nineteenth-century critics and reviewers championed unindebted originality as separate from, and superior to, the era's sometimes more obviously intertextual scientific literature. Significantly, these changes in critical expectations for autonomous creation in literature developed in connection with, for example, ruptures between religion and science, a pervading conservatism in British society following the French Revolution, and the increasing separate professionalizations of literature and science.[21]

These arguments remain important for *Regenerating Romanticism* because they establish that—despite nineteenth-century critical efforts to distinguish, for instance, the "literature of knowledge" from the "literature of power"—both writers of scientific literature and now-canonical male Romantic-era authors (each of whom in some way also absorbed science into their poetry), in reality, often sought to achieve literary originality that incorporated both novelty and intertextuality; this recognition helps to explode long-standing justifications for precluding scientific literature from the canon. Nevertheless, while the women writers of scientific literature whom I explored in *Questioning Nature* generally sought to convey novelty as well as the lasting achievement of originality, Darwin self-deprecatingly claimed for his poetry only "novelty" (which could be interpreted as the merely fashionable and fleeting, as opposed to the more enduring quality of originality) and trivialized his verse. Darwin thereby inadvertently set forth the terms by which some early nineteenth-century critics and major male Romantic poets would denigrate his work, and (inaccurately) scientific literature more broadly. It is a tricky dynamic because, while other writers of scientific literature lauded the originality of Darwin's poetry they, like the antagonizing critics, reviewers, and male Romantic poets, also often condemned his scientific poetry as lacking sensibility.

In fact, in several important ways, Darwin's *Botanic Garden* counters the poetic methods suggested in Aikin's *Essay on the Application of Natural History to Poetry*, which many writers recognized as providing guidelines for what the larger movement of scientific literature sought to achieve. First, although Aikin became more interested in botanical studies shortly following the publication of his *Essay*, his text favors zoology over botany. Dedicating the *Essay* to the British zoologist Thomas Pennant, he recommends that poets versify zoological species as having in common with humanity "somewhat of a moral and intellectual character" and discourages botany due to the difficulty of communicating "color, scent, and taste" in verse.[22] Of course, Darwin would focus his most successful poems on botany. Second, Aikin discounts attempts to teach systems through poetry.[23] Yet Darwin specifically versifies the Linnaean system of botanical classification. Third, while Aikin describes poets' engagements with natural history as a means to "original genius" or originality as well as novelty, as I have mentioned, Darwin solely emphasizes "novelty" (not "originality") as fundamental to his poetic methodology in his prose interludes in *The Loves of the Plants*. Fourth, and perhaps most important for this study, Aikin recommends versifying natural objects that would appeal to the reader's sensibility and "moral reflection."[24] For example, he suggests to poets the subject of decoy ducks that hunters have "taught to be the crafty betrayers" of wild ducks, who then find "themselves unexpectedly entrapped," providing poet-naturalists with "copious matter as well for sentiment as description."[25] Conversely, in Aikin's brief 1815 biographical essay about Darwin, with whom he corresponded during the latter's lifetime, he writes that "Darwin was in feature and person coarse and uncouth," and his "disposition was humane, though with *little sensibility*."[26] Additionally, Aikin explains that while *The Loves of the Plants* initially achieved great success, *The Economy of Vegetation* proved "less amusing to common readers" due to "the more scientific nature of its subject . . . the peculiarities of the writer's manner, and the want of *human interest* in the design."[27]

Thus, although Anna Seward declared in her published letters that Darwin's *Botanic Garden* fulfilled Aikin's recommended "union of natural history . . . with poetry," Darwin also significantly departs from Aikin's plan.[28] This is not to say that other authors who were blending literature and natural history did not also sometimes diverge from Aikin's treatise; they did, as the era's extensive literature on botany, for instance, demonstrates.[29] However, especially in Darwin's verbal emphasis on "novelty" to the exclusion of "originality," and his scientific poetry's lack of sensibility as perceived by many of his contemporaries, he became vulnerable to devastating critiques that then negatively and unjustly affected the texts of other authors of scientific literature as well. By 1815, in Aikin's words, the popularity of Darwin's *Botanic Garden* had "faded away, and it seems doubtful whether it will retain a place among the approved productions of the

British Muses."[30] My comparisons between Aikin and Darwin reinforce the point that Darwin cannot be viewed as representative of other writers who combined literature and natural history, many of whom, like Aikin, also critiqued aspects of Darwin's work. When Darwin's poetry is decentralized in this way, a clearer and more authentic understanding of the dynamics involved in this movement of scientific literature, and the goals and challenges of individual participating writers, becomes more apparent.

Sensibility and Romanticism

As the relationship between sensibility and scientific literature represents the central category of inquiry in this book, and sensibility is the quality that so many writers accused Darwin and his scientific poetry of lacking, it is integral to understand the concept's complexities in this era. The literature of sensibility forms a historical tradition stretching back to the late seventeenth century and is sometimes viewed as lasting through the 1840s and beyond.[31] Within this continuum, sensibility was understood and performed in multiple ways in England as, for instance, civility and politeness, refined moral virtue, compassion and sympathy, and the restraint of passion or bodily desire.[32] It spanned concerns with ethical, aesthetic, and physiological qualities and, through "the individual's immediate sensory and affective experience, becomes the basis for a new empiricist epistemology and a new model of moral agency."[33] However, as Jennifer Keith observes, "some readers in the eighteenth century and today have questioned whether the emotions stimulated by sensibility are either positive or socially efficacious," as they sometimes may be viewed as "a narcissistic exercise, enabling the reader who sympathizes to feel morally and aesthetically superior to the sufferer and to anyone incapable of feeling such sympathy."[34] Indeed, according to John Mullan, the effects of sensibility are so ambivalent that its ideological functions may be difficult to determine.[35] This difficulty also exists in attempts to identify sensibility with a particular gender, as "sensibility bestowed a quality previously seen as 'feminine' upon men and established men as the best practitioners of sympathy" because it was argued that men "had greater rational powers to balance and shape these impulses of the heart."[36] While G. J. Barker-Benfield examines sensibility as a "feminizing" influence, Claudia L. Johnson clarifies that characteristics of sensibility "are valued not because they are understood as feminine, but precisely and only insofar as they have been recoded as masculine."[37] Thus, feminist writers of the late eighteenth century, Mary Wollstonecraft and Hannah More, each embraced the concept of "rational sensibility" in which "one should think feelingly, but never let one's feelings annihilate one's capacity for discerning the truth."[38]

These writers' assertions of "rational sensibility" therefore extend the more general understanding of sensibility as "the ability to feel what others feel" and, interestingly, link with some scholars' differentiations between sentiment and sensibility.[39] According to Ann Jessie Van Sant, "*Sensibility* and *sentimental* are in one respect easy to separate: sensibility is associated with the body, sentiment with the mind. The first is based on physical sensitivity and the processes of sensation; the second refers to a refinement of thought."[40] G. S. Rousseau further qualifies that "sensibility was the larger of the two, touching almost every aspect of life; sentimentalism came later, especially in imaginative literature, and was the more religious, moral, literary, and far less aristocratic."[41] However, other scholars, including G. J. Barker-Benfield and Markman Ellis, contend that the terms are so closely allied that writers often used them interchangeably.[42] As Stephen Ahern points out, such discrepancies encouraged parodies of sensibility alongside its more serious expressions and debates, displaying that its "contradictory cultural discourse [was] rooted in an unstable complex of assumptions about the ontological status and political implications of social identity," so that "by 1800 the voices against sentiment were as clamorous as those for it."[43]

Significantly, various scholars have asserted that the now-canonical male Romantic-era authors appropriated this tradition of sensibility in ways that proved to women writers' detriment. Alan Richardson argues that these "male writers drew on memories and fantasies of identification with the mother in order to colonize the conventionally feminine domain of sensibility."[44] Stuart Curran contends that the culture of sensibility "was largely a female creation," constituting "a central concern in writing by women," as well as "the foundation on which Romanticism was reared."[45] James Chandler has recently provided insight into this question of sensibility, explaining that scholars of Romanticism in the 1960s "tended to reinforce the idea that Romanticism had left sensibility behind," so that "sensibility is a major blind spot in" the literary criticism of, for instance, M. H. Abrams.[46] Analyzing women's "introspective" and "self-reflexive" poetry of sensibility, Curran further suggests that "it is women who truly do speak 'the real language of men,'" noting that William Wordsworth, for example, adopted Charlotte Smith's "style, the long, sinuous verse paragraphs, the weighted monosyllables, the quick evocations of natural detail;

in matter, the absorbing and self-mythicizing voice and the creatures of its contemplation—the aged, the idiots, the female vagrants, the exiled and alienated."[47] Curran also references Wordsworth's often-quoted, but often-shortened, 1833 description of Smith as "a lady to whom English verse is under greater obligations than are likely to be either acknowledged or remembered. She wrote little, and that little unambitiously, but with true feeling for rural nature, at a time when nature was not much regarded by English Poets; for in point of time, her earlier writings preceded, I believe, those of Cowper and Burns."[48] Wordsworth makes these remarks in a footnote to his poem, "Stanzas ... Off St. Bees' Head," admitting that the "form of stanza in the poem, and something in the style of versification, are adopted from" Smith's poem "St. Monica." Although Curran calls this subtextual acknowledgment "generous praise," Jacqueline Labbe notices Wordsworth's "offhand[ed] tone."[49] In fact, Wordsworth here downplays Smith's importance and the volume of her work ("She wrote little, and that little unambitiously") and, as he does elsewhere, ignores the presence of so many other authors of sensibility who "wrote with true feeling for rural nature" in the decades preceding publication of his own works.[50] Thus, rather than expressing generosity, it is tempting to hear self-justification and a twinge of guilt for his part in ensuring that Smith's poems of sensibility and the movement of scientific literature in which she played a central role would not be historically "acknowledged or remembered."

In this vein, Anne Mellor incisively writes of the fate of sensibility, "When he claimed that poetry is 'the spontaneous overflow of powerful feelings' from a man who has 'thought long and deeply,' Wordsworth and the Romantic poets who followed him effectively stole from women their primary cultural authority as the experts in delicate, tender feelings and, by extension, moral purity and goodness.... After these canonized Romantic poets had stolen their emotions, their intuition, their capacity for imagination and fancy, their romances, and their affinity with nature, what did women have left?"[51] Mellor further states that, "in masculine Romanticism, we often see the poet appropriating whatever of the feminine he deems valuable and then consigning the rest either to silence or to the category of evil."[52] While agreeing with Mellor, I also would argue that if Wordsworth and some of the other now-canonical male Romantic-era writers had merely "stolen" sensibility, then perhaps, as in cases of plagiarism, it would have been easier to locate, condemn, or correct long ago; instead, as I further explore in my conclusion, their selective appropriation and manipulation of sensibility's socioideological functions served to justify the creation of what became—especially through literary critics' and reviewers' nineteenth-century constructions of the canon—an often sexist, racist, and exclusionary system of literature, especially poetry. Wordsworth defined poetry in 1800 as "the history or science of feelings," seeking to install "poetry as a mode of both social and aesthetic inquiry in its own right."[53] Yet, in his 1802 preface to *Lyrical Ballads*, he separates the natural sciences from sensibility or feeling so that the scientific literature of sensibility by writers including Charlotte Smith becomes a paradox that according to this (Wordsworthian) version of Romantic-era aesthetics cannot, and therefore subsequently did not, canonically exist—except, of course, in the ways in which it was appropriated by the now-canonical male Romantic poets.

Romanticism, Wordsworth's Preface, and Darwin

Despite renewed interest in Erasmus Darwin and his works in recent decades, some scholars paint a bleak picture of his poetic legacy, stating, for instance, that his verses "characterize" a "degenerate intellectual and poetic tradition that taught the Romantics ... as much through negative as through positive example," and that he "has been universally judged to be a remarkably bad poet."[54] Seeking to explain how Darwin's *Botanic Garden* could achieve such success in the 1790s and quickly thereafter fall into relative obscurity, his grandson, Charles, writes in his *Life of Erasmus Darwin* (1879) that "no one of the present generation reads, as it appears, a single line of it," and, while "the downfall of his fame as a poet was chiefly caused by the publication of the well-known parody the 'Loves of the Triangles'" in the *Anti-Jacobin* (1798), "public taste was at this time changing."[55] Charles supposes that "it was generally acknowledged, under the guidance of Wordsworth and Coleridge, that poetry was chiefly concerned with the feelings and deeper workings of the mind; whereas Darwin maintained that poetry ought chiefly to confine itself to the word-painting of visible objects."[56] Charles's phrasing is striking in retrospectively attributing authoritative "guidance" to Wordsworth and Coleridge in effecting a "chang[e]" in "public taste" toward poetry "concerned with feelings and deeper workings of the mind" as opposed to Darwin's emphasis on "visible objects" in his own poetic theory. However, the problem here, as I have already begun to show, is that not only had poets of sensibility been writing verse "concerned with the feelings and deeper workings of the mind" for decades before Wordsworth and Coleridge published their poetry, but some of the era's poets who combined sensibility and science in their works also critiqued Darwin's emphasis on the "eye," as opposed to the heart. Thus, in 1879, Charles unwittingly repeats the erroneous narratives that had already rewritten history and that were put forth by Wordsworth and Coleridge, and later propounded especially by critics and reviewers of the 1820s and 1830s, establishing conceptions of now-canonical Romantic-era aesthetics; such narratives not only elide earlier and concurrent writers of sensibility but also continue to encourage today's scholars to wrongly present Darwin as "characteriz[ing]" the era's scientific literature rather than acknowledging its diversity and complexity.

Since Darwin already enjoyed a thriving career as a physician and the reputation of a natural philosopher within circles of like-minded men in, for instance, the Lunar Society of Birmingham, he expressed concern that becoming a published poet might affect his medical practice. His reticence in this regard caused him to initially publish *The Loves of the Plants* anonymously, noting that "the addition of my name would seem as if I thought it a work of consequence," and to downplay his seriousness as a poet even within the poem itself.[57] In Darwin's preface to this work, he acts as a showman, describing the poem as a camera obscura, and invites readers who are "perfectly at leisure for such a trivial amusement [to] walk in, and view the wonders of my enchanted garden," explaining that it also may be "contemplate[d] as diverse little pictures suspended over the chimney of a Lady's dressing-room."[58] Through prose "interludes" that divide the poem's cantos and record conversation between the Poet and his Bookseller, Darwin asserts that the difference between poetry and prose is that "the Poet writes principally to the eye," and that words expressive of "ideas derived from visible objects" comprise "the principal part of poetic language," while "the Prose-writer uses more abstracted terms." Thus, he justifies his use of footnotes, writing that "Science is best delivered in Prose."[59] He also repeatedly emphasizes the importance in poetry of expressing "sublimity, beauty, or novelty ... in picturesque language, so as to bring the scenery before [the reader's] eye."[60] Moreover, he explains, in a statement that seems to counter the aesthetic goals of Aikin and most writers of scientific literature who closely examine the natural world, that "the further the artist recedes from nature, the greater novelty he is likely to produce."[61]

Darwin's *Botanic Garden*, and especially *Loves of the Plants*, initially brought recognition for his originality. In the mid-1790s, Coleridge called Darwin "the first literary character in Europe, and the most original-minded Man," enthusing that he had "a greater range of knowledge than any other man in Europe," as well as being "the most inventive of philosophical men."[62] Anna Seward claimed that "not one great Poet of England is more original than Darwin" and hailed his poem as "form[ing] a new class in poetry," while William Cowper wrote that Darwin's "style and manner are unquestionably his own."[63] Richard Edgeworth also stated that the poem "has silenced for ever the complaints of poets, who lament that Homer, Milton, Shakespeare and a few Classics had left nothing new to describe."[64] When Darwin then published his medical treatise *Zoonomia* (1794–96), Thomas Beddoes thought it "perhaps the most original work ever composed by mortal man," which would "place the Author among the greatest of mankind, the founders of sciences."[65] Looking back at the first half of the 1790s, Wordsworth admitted, "my taste and natural tendencies were under an injurious influence from the dazzling manner of Darwin"; and this "dazzling manner" soon drew complaint.[66]

Thomas James Mathias's poem *The Pursuits of Literature* (1794) provides one of the first public critiques of Darwin's verses. Accusing Darwin's work of promoting sexual libertinism, Mathias expresses particular concern for "how young ladies are instructed in terms of botany" and satirizes *The Loves of the Plants* by hyperbolizing its bawdiness:

> On the luxurious lap of Flora thrown,
> On beds of yielding vegetable down,
> Raise lust in pinks; and with unhallow'd fire
> Bid the soft virgin violet expire.[67]

In a footnote, Mathias describes Darwin's *Botanic Garden* as "glittering words" that "sacrifice propriety and just imagery to the rage of mere novelty."[68] Mathias's assertion that Darwin's poetry represents "mere novelty," echoing Darwin's use of the term "novelty" (rather than "originality") in his poetic theory in *The Loves of the Plants*, became a frequent criticism. Of course, the insinuation in such critiques is that literature providing "mere novelty" focuses only on the new in the sense of the latest fad, or something that will not last. Sensing the effectiveness of this censure, Wordsworth repeatedly declares in his preface to *Lyrical Ballads* that his own "class of Poetry" is "well adapted to interest mankind *permanently*" while also contributing to its "moral relations" (615). Wordsworth again jabs at Darwin by claiming to offer a remedy for "the present outcry against the *triviality and meanness*, both of thought and language, which some of my contemporaries have occasionally introduced into their metrical compositions"

(597). He writes that this is "the language of men who speak of what they do not understand; who talk of Poetry as a matter of amusement and idle pleasure" (604). Further contrasting his poetic method with Darwin's, Wordsworth suggests that poetry should not teach readers new information, for the Poet must give "immediate pleasure to a human Being possessed of that information which may be expected from him, not as a . . . natural philosopher, but as a Man" (605). For him, "Poetry is the first and last of all knowledge—it is as immortal as the heart of man" (606).

Too late, Darwin seems to have realized that his emphasis on novelty in his poetic theory, aligning his verse with novelty's dual meanings of the new (which, of course, could have positive connotations for literature) and the trivial, allowed some contemporary critics and writers to deem his work as insignificant and short lived. In an exasperated note to *The Temple of Nature* (1803), Darwin defends the importance of novelty and subtly hits back at his critics who did not appreciate this quality of his verse, explaining that "the pleasure of novelty is produced . . . in comparing uncommon objects with those which are more usually exhibited" and thus "is less perceived by . . . very ignorant or very stupid people, or by brute animals"; for, to understand "objects of Taste requires some previous knowledge of such kinds of objects, and some degree of mental exertion."[69] He even asserts that the lack of pleasure in novelty, or "new trains of ideas," is "one cause of the torpor of old age, and of death."[70] Yet, despite the denigration of novelty in certain texts of Romantic-era aesthetics, including De Quincey's distinction of the "literature of knowledge" from the "literature of power," it should also be remembered that novelty was integral to the poetic philosophy of Wordsworth and Coleridge. In *Biographia Literaria* (1817), Coleridge recounts that, while planning the *Lyrical Ballads*, he and Wordsworth agreed "on the two cardinal points of poetry, the power of exciting the *sympathy* of the reader by a faithful adherence to the truth of nature, and the power of giving the interest of *novelty*, by the modifying colors of imagination."[71] Indeed, Wordsworth's poetic task was "to give the charm of novelty to things of every day" and thereby awaken "the mind's attention from the lethargy of custom" and "familiarity."[72] Yet, at the same time, this commitment to novelty in the familiar did not prevent Wordsworth from, for example, adding footnotes to his poem "Ruth" (1800) that provide Linnaeus's scientific name for the magnolia, *Magnolia grandiflora*, and that reference other flowers from William Bartram's *Travels through North and South Carolina, Georgia, East and West Florida, the Cherokee Country, the Extensive Territories of the Muscogulges, or Creek Confederacy, and the Country of the Chactaws* (1791).[73] Thus, novelty represents another poetic quality that can be variably appropriated, valued, and disparaged in canonical Romantic aesthetics, depending on the writer who wields it.

Significantly, in addition to condemning Darwin's novelty, Wordsworth and Coleridge denounce him as failing in their other "cardinal [point] of poetry": sympathy, or sensibility. In *Biographia Literaria*, Coleridge compares "Darwin's work to the Russian palace of ice, glittering, cold and transitory."[74] Conveying Coleridge's distaste for Darwin's "glittering" language as well as "transitory" novelty and poetic fame, this comparison of Darwin's verse with "cold" and "ice" crucially registers an accusation that this older poet's work lacks warmth, passion, sympathy, emotion. In a series of lectures on literature delivered in 1811–12, Coleridge declared that "poetry is opposed to science."[75] Of course, Coleridge's description of Darwin's poetry as unfeeling also recalls Wordsworth's claim that "all good poetry is the spontaneous overflow of powerful feelings," and his definition of the Poet as a man "with more lively sensibility, more enthusiasm and tenderness, who has a greater knowledge of human nature, and a more comprehensive soul, than are supposed to be common among mankind" (598, 603). Similarly, Wordsworth's poem "A Poet's Epitaph" (1800) includes a stanza with Darwin's *insensibility* in mind, scoffing,

> Physician art thou? One all eyes,
> [Natural] Philosopher! a fingering slave,
> One that would peep and botanize
> Upon his mother's grave?[76]

Regarding the second line of this excerpt, Nicola Trott observes that Wordsworth "added a certain incestuous insult to the injury."[77] Moreover, Darwin's emphasis on the importance of the eye in composing poetry here becomes uncomfortably indicative of a voyeur. Indeed, Wordsworth implies that Darwin is "all eyes," and thus no heart, no human sympathy or feeling. Yet, five years later, it was actually Wordsworth who botanized on a family member's imagined or symbolic gravesite in his "Elegiac Verses, in Memory of My Brother, John Wordsworth" (comp. 1805, pub. 1842), after his brother drowned in a shipwreck. Proposing to place a "monumental Stone" on the spot where he and his brother last parted, Wordsworth perceives a flower possessing a "multitude of purple eyes" that seem to stare back as Wordsworth repeats that he "see[s] it," thus interestingly amplifying Darwin's stress on visual objects while Wordsworth also identifies the plant in a note as "the Moss Campion (*Silene acaulis*, of Linnaeus)."[78]

Still, an 1800 article in the *Anti-Jacobin Review and Magazine* demeans Darwin's verses in a survey of "didactic" poems deemed "dry and uninteresting"; the reviewer urges that "literary men should always remember—that, 'in order to write successfully, we should feel vividly,'" and thus the poet should write "to us from his immediate feelings."[79] Underscoring this point, the reviewer explains that the subject of Darwin's *Botanic Garden* is "ill-chosen," for, "how is it possible to enter into the feelings of plants?"[80] He therefore declares that Darwin's poetry "is only the fashion of the day . . . whilst the principles of taste are immutable," so that "it becomes every lover of the Muse to watch the inroads of science with an eye of jealousy . . . [and] to check her influence, lest the intermixture of scientific discovery with poetic invention should become fashionable, and every spark of poetry at length be quenched in the phlegm of philosophy."[81] As Noel Jackson puts it, "The path that lay beyond this utterance runs straight through many of the major statements of Romantic and early Victorian literary aesthetics: Wordsworth's distinction between the poet and the man of science in the 1802 preface, Coleridge's definition of poetry as the antithesis of science, William Hazlitt's (or Peacock's, or Keats's, etc.) vision of natural philosophy as hostile to the spirit of poetry, Thomas De Quincey's distinction between the literature of knowledge and the literature of power."[82]

However, fascinatingly, in multiple ways, Darwin's works also may be viewed as representative of sensibility. Not only do poets arguably demonstrate a greater capacity for sensibility by "enter[ing] into the feelings of plants" than, say, those of humans, but Darwin also establishes important connections with sensibility through his use of personification, "ideal presence," and "nerve theory" (as I will discuss below). Darwin's personifications of individual flowers and plants, as well as the Goddess of Botany and objects such as a compass and thermometer, comprise the central conceit of *The Botanic Garden*. In eighteenth-century poems of sensibility, personifications provided powerful stimulants to sensory experience.[83] In *Elements of Criticism* (1762), Henry Home, Lord Kames describes these intense sensory responses to poetic imagery, especially personification, as "ideal presence," a concept referenced by Darwin in his first prose interlude to *The Loves of the Plants*. According to Kames, "ideal presence" produces almost a state of hallucination, creating the impression for the reader of being an eyewitness to the incidents that are poetically portrayed. As Eric

Rothstein suggests, this imaginative response to personification underscored "a certain social kinship" among readers that countered the focus on individual emotional experience, engendering a sense of community and connection.[84] In fact, Keith argues that, by extending a realm of private experience to the public and putting "flesh and blood on what could otherwise seem empty abstractions, . . . personification *is* sensibility."[85]

In Wordsworth's 1802 preface, he claims to employ personifications as "a figure of speech occasionally prompted by passion" and "utterly to reject them as a mechanical device of style," wishing "to keep the Reader in the company of flesh and blood" (600). Thus, Wordsworth does not object to personification per se; in his poem "Lines Written in Early Spring" (1798), for instance, he declares, "And 'tis my faith that every flower / Enjoys the air it breathes." Later in the preface, Wordsworth conjoins "Science" and personification, famously writing:

> If the labors of Men of Science should ever create any material revolution, direct or indirect, in our condition, and in the impressions which we habitually receive, the Poet will sleep no more than at present, but he will be ready to follow the steps of the Man of Science, not only in those general indirect effects, but he will be at his side, carrying sensation into the midst of the objects of the Science itself. The remotest discoveries of the Chemist, the Botanist, or Mineralogist, will be as proper objects of the Poet's art as any upon which it can be employed, if the time should ever come when these things shall be familiar to us, and the relations under which they are contemplated by the followers of these respective Sciences shall be manifestly and palpably material to us as enjoying and suffering beings. If the time should ever come when what is now called Science, thus familiarized to men, shall be ready to put on, as it were, a form of flesh and blood, the Poet will lend his divine spirit to aid the transfiguration, and will welcome the Being thus produced, as a dear and genuine inmate of the household of man. (606–7)

Of course, despite Wordsworth's projection of this melding of poetry and science as occurring at some future time, numerous poets had already incorporated various fields of the natural sciences, particularly botany, into their verse. His series of "if" clauses deny the existence of that earlier scientific poetry, or at least its legitimacy as poetry. Moreover, Wordsworth's offer that the poet may transfigure "Science" and the "objects of science" into "flesh and blood" appears to propose personifications of the kind that had already materialized in the botanical poetry of, for instance, Erasmus Darwin and Charlotte Smith. In this way, he not only elides the existence of previous scientific poems and their communication of changes that the sciences had already effected in "the impressions which we habitually receive," but also separates those poems' use of personification from sensibility, seemingly indicating that what scientific poetry requires, and has not yet achieved, is "sensation" and "familiar[ity]" of subject matter. Nevertheless, Linnaean botany was extraordinarily popular and easily considered "familiar" by this time. As scholars such as Catherine Packham have suggested, Wordsworth's more pressing critique thus appears to be leveled at the lack of "sensation" in scientific poetry.[86] Packham argues that "Darwin's personification does not, as it does for Kames and Blair, derive from poetic passion, but rather offers a sustained joke for the amusement (and education) of its readers."[87] This perception of Darwin's "detached wit" encouraged many of his contemporaries to critique his verse as lacking sensation, even as his engagements with personification, ideal presence, and nerve theory intellectually—if not always emotionally—place him directly in the realm of sensibility. However, by not naming Darwin specifically, Wordsworth's censure appears to be directed at scientific poetry more generally and thus erases the fact that in the previous decades many male and female poets had already incorporated emotional registers of sensibility within their scientific poems.

In fact, as I have been arguing, numerous authors who blatantly combined literature and natural history, such as Charlotte Smith, Anna Seward, Helen Maria Williams, Sydney Owenson, and Anna Barbauld, wrote specifically within the tradition of sensibility. As I show in chapters 2, 5, and 6 of this study, women such as Seward and Smith also critique Darwin's botanical verses as lacking sensibility; nevertheless, I additionally demonstrate in chapter 3 that Darwin published other verses that fit easily within more emotional conceptions of sensibility, displaying that his oeuvre's complexity should be acknowledged. Chapters 5 and 6 exhibit that each of these women, too, censured Wordsworth's poetry and poetic methods; and, as many scholars have discussed, they also critiqued one another. All of the female writers listed above considered themselves to be serious poets and aspired to a lasting notion of originality for their work, placing faith in posterity for their texts' permanence. Seward, for instance, wrote, "I have always destroyed every little production of my own, if, on revising it, after the effervescence of composition had subsided, I could not find that it contained something original, either in the thoughts themselves, or in their combination"; and Smith viewed her poetry as her most serious and lasting artistic achievement.[88] Some of these women even sought to rectify Darwin's self-deprecations by imputing originality and permanence to his work, as when Seward claims that "Darwin's excellence consists" in the "perfect originality of his [poetic] pictures" in *The Botanic Garden*, "which secure its immortality."[89] Thus, I argue that the traditional narrative regarding the fate of Darwin's work due to his use of novelty and suggested lack of sensibility not only is incomplete and inaccurate, but also became the "straw man" enabling the historical exclusion from the canon of the larger movement of scientific literature, which was much more diverse in its literary practices and substantially composed of women writers. Further evidence of this straw man tactic can be found in numerous now-canonical texts of Romantic-era aesthetics, but it is useful here to more closely analyze Wordsworth's preface. And, again, Coleridge and other critics and reviewers identified the preface as a direct effort to repudiate Darwin, further contributing to the means by which individual authors of scientific literature were categorized and dismissed as belonging to the supposed "Darwinian school" despite their various differences.

Significantly, Wordsworth's differentiations between literature and science in the preface appear fraught with contradictions. Indeed, although Coleridge thought it "impossible to deny the presence of original genius" in Wordsworth's poetic theory, he also explains that "with many parts of the preface [to *Lyrical Ballads*] ... I never concurred; but on the contrary objected to them as erroneous in principle, and as contradictory."[90] So, for example, despite attempting to draw distinctions between literature and science, Wordsworth writes that "the Man of science seeks truth," and the Poet "rejoices in the presence of truth," prompting Coleridge to offer to clarify that "a poem ... is opposed to works of science, by proposing for its *immediate* object pleasure, not truth."[91] And it is not difficult to find more of the contradictions to which Coleridge alludes. For example, Wordsworth asserts that the Poet's knowledge "cleaves to us as a necessary part of our existence, our natural and unalienable inheritance," and differentiates the knowledge of the "Man of Science" as a "personal and individual acquisition, slow to come to us, and by no habitual and direct sympathy connecting us with our fellow-beings" (606). However, later in the preface, he states that "an *accurate* taste in poetry, and in all the other arts ... is an *acquired* talent, which can only be produced by thought, and a long-continued intercourse with the best models of composition" (614). If "accurate taste in poetry" is an "acquired talent" requiring "long-continued" thought and study of the "best models," then how can the knowledge imparted through poets also be "our natural and unalienable inheritance"? If poetic knowledge must be learned or "acquired," then it cannot be understood in essentialist terms as being inherent, "natural," or "inherit[ed]," and thus does not seem to differ from the study noted as being required to attain knowledge of science.

Additionally, Wordsworth's suggestion that scientific knowledge has "no habitual and direct sympathy connecting us with our fellow-beings" seems strange, especially in his application of this to the "Botanist," as the fields of natural history around the turn of the century were generally considered an "open and egalitarian" form of natural inquiry. Indeed, writers of scientific literature often incorporated themes of sympathy and sensibility, and contemporaries of the French Revolution sometimes represented natural history as science's democratic ideal "connecting us with our fellow beings." In Britain and Continental Europe, the natural sciences thrived through national and international networks of salons, scientific societies, correspondence, collaboration, and the widespread involvement of both observant amateurs and "professed naturalists."[92] Moreover, while in the preface Wordsworth writes that "the Poet binds together by passion and knowledge the vast empire of human society" and instead relegates the "Man of Science" to "solitude," in his "Essay, Supplementary to the Preface" (1815), he incongruously claims that the poet's "grand thoughts . . . are most naturally and most fitly conceived in solitude" (606, 661). To say the least, Wordsworth appears confused as to how he might distinguish the Poet from the "Man of Science."

Crucially, Wordsworth's preface also interconnects gender and science. As other scholars have discussed, and any modern reader must notice, Wordsworth's theorization that poetry should be written in "the real language of men," and his description of the Poet as "a man speaking to men," misogynistically omits the many women poets, readers, and critics of the era. Nevertheless, Lucy Aikin herself (the daughter of John Aikin) reviewed Wordsworth's 1807 *Poems* in the *Annual Review*, and, although she "rather begrudgingly" acknowledges that he "doubtless possesses a reflecting mind and a feeling heart," she accuses him of many of the failings that he and other reviewers had sought to impute to Darwin. She writes, "Nature seems to have bestowed on [Wordsworth] little of the fancy of a poet, and a *foolish theory* deters him from displaying even that little. In addition to this, he appears to us to starve his mind in *solitude*.—Hence the undue importance he attaches to *trivial* incidents . . . and hence, finally, the unfortunate habit he has acquired of attaching exquisite emotions to objects which *excite none* in any other human breast."[93] Of course, in his preface, Wordsworth additionally sexes the "Man of Science" despite the fact that the natural sciences, and especially botany, were often studied and practiced by women; and, considering that many of the poets versifying subjects of natural history also were women, his frequent masculinizations of both poetry and science seem calculated to exclude women from recognition or participation in each of these fields, let alone from combining them within their writing. And, finally, it is remarkable that even as Wordsworth insists that poets employ "the real language of men" and ostensibly seeks to preclude science from the realm of poetry, he repeatedly makes his case in glaringly scientific, and especially botanical, terms. For example, he writes that "the manners of rural life *germinate* from those elementary feelings," and "essential passions of the heart *find a better soil* in which they can attain their maturity," and he asks, "Why trouble yourself about the *species* till you have previously decided upon the *genus?*" (597, 613). Thus, Wordsworth clearly struggled in, to borrow Coleridge's terms, his "erroneous" and "contradictory" efforts to separate poetry from science, and particularly from botany.

Linnaean Botany in the Romantic Era

Scientific, including botanical, discourse stretching back to the mid-seventeenth century provides important contexts for understanding sensibility. While sensibility etymologically refers to a physiological capacity of sensation or sense perception, medical writers of the era also used it "to connote self-consciousness and self-awareness."[94] These medical and anatomical notions of sensibility gained currency especially through conceptions of "nerve theory"; this theory developed out of assertions by physicians such as Thomas Willis, who viewed the brain as the seat of the soul and the nerves as its only source of knowledge input, as well as by John Locke, who claimed that all knowledge derives from sensory experience and that nerves control human consciousness.[95] As Jessica Riskin has shown through her examination of French Enlightenment science, which additionally influenced English thought, "sensibility transformed the practice of the sciences as well as the arts and literature"; in fact, sentimental empiricists' jeremiads challenging instances of "chilly, arid rationalism" within the sciences ironically proved so convincing that they "erased their own traces and wrote themselves out of history," leaving the impression that the Enlightenment sciences were devoid of sensibility when this was not the case.[96] Botanists, too, incorporated these scientific concerns of sensibility into their conceptions of plant function and anatomy, especially in studies of, for example, the sensitive plant. Despite accusations of Erasmus Darwin's lack of sensibility from contemporary poets, literary critics, and reviewers, such scientific and intellectual contexts for sensibility become especially apparent in his prose and poetry, where he conjoins his medical and botanical knowledge to explore plant physiology's analogies with human anatomy and consciousness. Moreover, botany was the most popular field of the natural sciences during this period, especially among women, and many writers employed plant studies as a mode of contemplating and conveying the various emotional and ideological functions of sensibility.

Much scholarship now exists exploring the botanical literature of this era, and foundational studies by literary critics and historians, including Ann B. Shteir, Londa Schiebinger, Alan Bewell, Sam George, and Theresa M. Kelley, inform my work.[97] Thanks largely to these scholars, most readers interested in this period are already familiar with the basics of Linnaeus's methods set out in his first major publication, *Systema naturae* (1735), as well as with some of their significance to discussions of women's education, writing, and sexuality. Linnaeus's sexual system of plant classification dominated botanical practice in England "from about 1750 to 1810 and, among amateurs and field naturalists, long after that."[98] Centering his taxonomy on flowers' reproductive parts, this Swedish naturalist categorized plants by using the number of (male) stamens and their relative proportions to divide the vegetable kingdom into twenty-three classes, plus one class for nonflowering plants.[99] He employed social metaphors to describe the "nuptials" or "marriages" of plants, naming classes based on the root *andria* (the Greek term for "husband"), so that flowers with one stamen became Monandria, and those with two stamens became Diandria, and so on. Linnaeus then divided the classes into sixty-five orders based on the (female) pistils, and chose the root *gynia* ("wife") for that designation, so that flowers with one pistil became Monogynia. Recognizing that Linnaeus's binomial system of nomenclature and his classifications formed an easily practiced but "highly artificial" system, Schiebinger additionally points out that this naturalist's privileging of class over order incorporates traditional notions of gender hierarchy into the sciences by prioritizing the plants' male parts.[100] Similarly, Linnaeus represents the male part as active and the female part as passive.[101]

English translations of Linnaeus's botanical works began to appear in the 1750s; the *Gentleman's Magazine* lauded his classificatory system in 1754, and many magazines and periodicals contained discussions of botany in subsequent decades.[102] James Lee's Linnaean *Introduction to Botany*

(1760) became a "best-seller [that] remained a standard introductory work for fifty years."[103] While women traditionally had knowledge of plants as healers through medical botany or herbalizing and within domestic duties, after 1760 botany books were increasingly written and translated with female readers in mind. Jean-Jacques Rousseau drafted eight letters instructing women about botany, and Thomas Martyn translated these into English, adding notes and twenty-four letters of his own "Fully Explaining the System of Linnaeus" to create the influential *Letters on the Elements of Botany* (1785), which went through eight editions over the next thirty years. William Withering's *Botanical Arrangement of All the Vegetables Naturally Growing in Great Britain* (1776) became another important text "for at least a generation," and William and Dorothy Wordsworth acquired a copy of the third edition in 1800.[104] Withering's botany book incited controversy because he translated Linnaean terminology to exclude sexual reference, as he thought this would be more appropriate for a general audience, and especially for women. Darwin adamantly disagreed with Withering's decision to anglicize and desexualize Linnaeus's Latin terms, and the Botanical Society of Lichfield, of which Darwin was a member, published more literal translations of Linnaeus's *Systema Vegetabilium, A System of Vegetables* (1783) and *Genera Plantarum, The Families of Plants* (1787), consulting Samuel Johnson for advice on the former and thanking him in the preface. Darwin's versification of the Linnaean sexual system in *The Loves of the Plants* reflects botanists' portrayals of plants as erotic creatures, but it should be noted that James Grainger, William Whitehead, John Scott, Charlotte Smith, George Crabbe, and William Cowper, for example, all referenced Linnaean botany in their poetry prior to Darwin.

According to Shteir, by 1796 the sexualized and Latinized nomenclature was fully established, and, while public institutions of botany and science, such as the Royal Society and the Linnean Society, excluded women from formal participation, women were often encouraged to participate informally as plant collectors, illustrators, and students of plant taxonomy.[105] Frequent suggestions that botanical studies provided young women with healthful mental and physical exercise as well as opportunities to appreciate God in nature countered attacks on the impropriety of Linnaeus's sexual system and on "botanizing women" in, for instance, Mathias's *Pursuits of Literature* and Richard Polwhele's *Unsex'd Females* (1798). Perhaps reflecting on some of his own verses, Coleridge wrote in 1805 that he regretted "the sneers against women who study Botany."[106] In 1796, Priscilla Wakefield's *An Introduction to Botany, in a Series of Familiar Letters* constituted the first botany book written by a woman providing a systematic introduction to the science, and theorized that Linnaeus's "new system . . . may probably receive improvement from some future naturalist, [but] is never likely to be superseded."[107] However, especially during the 1810s through the 1830s, greater tensions slowly developed between popularizing and professionalizing impulses in the sciences, and increasing efforts were made through scientific specialization in the academies to masculinize (or "defeminize") botany. During this time, botanists also increasingly turned away from Linnaean methods and toward the "natural system," focused on a more comprehensive approach to morphological features of plants and natural affinities among plant groups in alignment with the classifications of Continental botanists including Antoine-Laurent de Jussieu and Augustin-Pyramus de Candolle. In 1828, the death of Sir James E. Smith, founder and president of the Linnean Society, marked the end of an era in British botany as John Lindley, who was appointed the first professor of botany at London University that same year, advocated the natural system and wrote prolifically about this approach to plant studies for various levels of audiences over the next four decades.

John Aikin earlier saw problems within Linnaean botany, as did Darwin, Charlotte Smith, and some of the other imaginative authors who employed this system. In Lucy Aikin's *Memoir of John Aikin* (1823), she writes that he became very interested in botany around the time of publishing his *Essay on the Application of Natural History to Poetry* (1777) and began recording in 1778 and for the next six years a calendar of "the names and habitats of all the plants [that] he observed . . . with the respective periods of flowering," which then informed his text "for young people, entitled *The Calendar of Nature*" (1784).[108] Nevertheless, she writes that Aikin questioned "the artificial system of arrangement laid down by Linnaeus" because the "natural methods of some preceding botanists possessed much superior attractions" as did "the classifications of Haller and Jussieu," but "the great convenience of the Linnaean system to the practical botanist reconciled him to its use."[109] Intriguingly, the simultaneous ease and problems of the Linnaean system arguably fit with Aikin's earlier encouragement for poets to explore or even seek to solve the various mysteries within the natural world that were *not* yet fully understood by naturalists, while also philosophizing more broadly on what *is* considered to be known.

Scholars have demonstrated the vast appeal of botany for British authors of the late eighteenth and early nineteenth centuries through articles and book chapters addressing not only each of the "big six" male Romantics but also writers including Jane Austen, Anna Barbauld, Harriet Beaufort, John Clare, George Crabbe, Mary Delany, Maria Edgeworth,

Sarah Fitton, Maria Graham, James Grainger, Sarah Hoare, Agnes Ibbetson, Maria Elizabeth Jacson, Elizabeth Kent, Jane Marcet, Elizabeth Moody, Henrietta Moriarty, Charlotte Murray, Sydney Owenson, Ann Radcliffe, Mary Roberts, Frances Arabella Rowden, John Scott, Anna Seward, Mary Shelley, Charlotte Smith, Robert Southey, Ann Taylor, Jane Taylor, Priscilla Wakefield, Helen Maria Williams, Mary Wollstonecraft, and Dorothy Wordsworth. Yet much more remains to be said, as we are just beginning to understand the importance of how and why imaginative writers employed botany and other fields of the natural sciences in this period. I hope that this study will help to reframe our knowledge of the era's movement of scientific literature and its diversity and complexity so that more thorough and accurate accounts of its history and practice, on its own terms, may be granted greater attention, analysis, and respect in broader scholarly discussions of Romanticism moving forward.

I also wish to note that, since scientific literature often lends itself to intermixing aspects of literary forms, in describing such instances I prefer concepts such as "mixture," "blending," and "hybridity," as terms and notions familiar to and applied by these authors, as in Darwin's discussions of hybridity in *The Loves of the Plants*.[110] I therefore avoid Wordsworth's categories of "composite species" or "composite order"—which, as Dahlia Porter explains, he altered to reference architectural ideas rather than science—as these would retroactively impose his terminology on the scientific literature of authors who became excluded from the literary canon due to the standardization of such aesthetics.[111] Thus, for example, although it may be tempting to describe the formal mixing that occurs in poems containing prose notes as being "like oil and water," it should also be remembered that many hybrid biological forms display the separate characteristics of each of its parent species yet still comprise a single organic being, just as those components of verse and prose in such cases still comprise a single text.[112] By exploring, where useful and appropriate, more of the concepts and terminologies actually employed by the writers of scientific literature, scholars can often better understand what each author sought to achieve through his or her own work.

Methods of Analysis: Chapter Descriptions

Regenerating Romanticism primarily investigates ways in which authors of scientific literature engaged with botany and sensibility during the late eighteenth and early nineteenth centuries. Its chapters provide insights into sensibility's significant role in negotiations of literature and science during the period in which these fields increasingly moved toward their separate professionalizations. Within the book's organizing scheme, my six chapters pair into three main thematic sections, respectively focusing on time, empire, and a category that I have designated as "in/effability," indicating notions of "description, classification, and defiance" that simultaneously employ and challenge scientific or taxonomic principles and practices. Although these three topics frequently overlap within my project, the chapter pairings provide a useful framework for charting authors' interconnections of botany and sensibility over the course of this period.

Highlighting understandings of seasonal temporalities, chapter 1 resolves the mystery of authorship for verses penned anonymously in a copy of the Lichfield Botanical Society's published translation of Linnaeus's *System of Vegetables* (1783), which has puzzled scholars of botanical literature for almost three decades. Setting these verses about botany and untimely weather in the context of other "Backwardness of Spring" poems, I argue that this unexplored subgenre of poetry responds to James Thomson's georgic *The Seasons* (1726–46); Thomson's poem engaged with botanical studies and contributed to establishing British conceptions of climate, time, sensibility, and "conjectural history," theorizing stages of society. While Thomson associates Britain with a "progressive" notion of seasonal time, he contrasts this experience with those of nations in the "torrid" and "frigid" zones, which he presents as functioning in a kind of atemporality because, in his poetic framework, they are confined to the seasons of summer and winter, respectively. Thomson's suggestion that Britons' capacity for sensibility depends on intertwining circumstances of climate and conjectural history influenced some subsequent poets, including the author of the anonymous manuscript botanical poem (which turns out not to be anonymous), Anna Barbauld, and John Clare, among others, to express varying levels of humor and concern when Britain experiences instances of seeming seasonal stagnation. Moving from these seasonal notions of time to daily, circadian rhythms, chapter 2 explores British poets' fascination with Linnaeus's theory of the floral clock in which certain flowers could be arranged in the shape of a clock enabling one to know the time of day (or night) simply by observing which flowers are open at that moment. While Erasmus Darwin's poetry portrays this phenomenon in more mechanical terms, describing flowers as possessing gears and clockwork, Charlotte Smith counters Darwin's mechanical and sexual imagery by instead portraying plants in terms of more active and vital agency through concepts of science and sensibility; later poets, including William Wordsworth and Felicia Hemans, refashion Linnaeus's floral clock theory to shift focus away from the botanical sciences and more pronouncedly to human feeling and internalized concerns with time and mortality.

Transitioning emphasis from temporality to empire, chapter 3 analyzes the works of the poet and travel writer Maria Riddell. I primarily discuss ways in which her *Voyages to the Madeira, and Leeward Caribbean Isles: With Sketches of the Natural History of These Islands* (1792, 1802) offers new understandings of mutability and hybridity in association with race, gender, sensibility, and nation in late eighteenth-century Britain and the British West Indian colonies. By foregrounding Riddell's engagements with natural history, this chapter places botanical and zoological classificatory systems in conversation with her descriptions of how white women colonists and enslaved African laborers utilized the natural world, while also displaying how she elicits British readers' sympathy or identification with these colonial inhabitants. I situate Riddell's works in relation to the ideas of contemporary naturalists, such as Linnaeus, Buffon, and William Smellie, as well as those of Erasmus Darwin, Edward Long, Janet Schaw, and John Clare. At a time of significant political revolutions, Riddell's travel narrative affords insight into implications for the Caribbean colonies and their function as an environment of sociobiological transformation. While eighteenth-century writing often represented the West Indies as a site of biological and social degeneration, this examination shows that writers such as Riddell depicted it also as a space of improvement and knowledge-making that produced hybrid identities, knowledge, and cultures. Significantly, the chapter concludes by focusing on two of Darwin's lesser-known poems that were published in Riddell's poetic collection *The Metrical Miscellany* (1802), displaying that, although Darwin and his *Botanic Garden* were sometimes critiqued as lacking feeling, he also wrote verses in more emotionally focused modes of sensibility. While maintaining concern with the West Indies, chapter 4 then argues that Sydney Owenson employs botanical imagery in her novel *The Wild Irish Girl* (1806) to critique important facets of British Empire, especially in regard to Irish colonialism and the 1800 Act of Union that formed the United Kingdom of Great Britain and Ireland. Portraying the Irish as metaphorical plants, as food for England's "body politic," Owenson seeks to awaken British readers' capacities for sensibility by gesturing toward this symbolic cannibalism as one of many forms of England's colonial consumption of Ireland displayed within her "national tale," and she highlights this notion further through comparisons between the Irish peasantry and enslaved African laborers in the Caribbean colonies. I show that, melding ideas of science and sensibility, as well as natural history and national history, Owenson subtly exposes England's exploitation of Ireland through the Act of Union.

Shifting from imperial to more local concerns, in my section on "in/effability," or "description, classification, and defiance," chapter 5 analyzes Charlotte Smith's final poem in *Conversations Introducing Poetry: Chiefly on Subjects of Natural History* (1804), "Flora," where she displays how particular locations reflect taxonomic systems' difficulties in placing ambiguity. Here, she depicts organic liminalities straddling between kingdoms, classes, orders, or species that set taxonomists on edge and alter her own claims to scientific knowledge as well. Revising the role of the poet-naturalist to align more closely with sensibility, Smith's sea exploration celebrates the natural world in the absence of strict placement within taxonomic orders as she sympathetically identifies with the unidentifiable, whose very existence retrospectively destabilizes her early confidence in the poem's botanical classifications. Through both her interrogations of male naturalists' and poets' portrayals of these botanical species and her efforts to relate feminine forms of political power as well as poetic creativity, Smith's emphases on sensibility and place suggest the potential for originality in revealing material mysteries and uncertainties within specific locations of the natural environment. From Smith's revisioning of taxonomical practices, chapter 6 then explores Anna Seward's efforts as a literary critic to reclaim or vindicate sensibility from negative associations in medical discourse that portrayed insanity as resulting from sensibility in excess. Fascinatingly, Seward seeks to restructure these perceptions of sensibility through her critiques of male poets' botanical verses. Examining her critical analysis of Darwin's *Botanic Garden* (1789, 1791) in her *Memoirs of the Life of Dr. Darwin* (1804), I demonstrate her previously unnoticed use of his theories and classifications of insanity from his medical text *Zoonomia* (1794–96) as well as sensibility's connections with nerve theory; through these ideologies, she suggests Darwin's own madness due to his poetry's *lack* of feeling or sensibility. Continuing to employ Darwin's medical theories, she additionally attributes madness to some of Wordsworth's botanical poetry and attacks his poetic theories as "dangerous" and egotistical, implying his failures to correctly express sensibility. Seward critiques similar qualities in the works of William Cowper, whose struggles with mental health were well known, and she does so within botanically related contexts that also interestingly connect with John Clare's distrust of Linnaean taxonomy.

Finally, the conclusion analyzes the fate of Romanticism and this era's movement of scientific literature in relation to the literary criticism and assertions of Hazlitt, Wordsworth, and especially De Quincey. These male writers each significantly influenced the formation of the literary canon; yet, as I show, they did so in ways that enforced exclusions, particularly based on gender and/or race. During the decades of the 1820s through the 1840s, which saw the increasing separate professionalizations of literature and science, these male authors sought to divide literature from scientific disciplines. However, I argue that, despite ostensibly asserting literature's rejection of and competition with the sciences in an attempt to consolidate literature's authority, these male literary critics in fact mirrored or adopted nineteenth-century scientific theories of biological sexism and (for De Quincey) racism, which influenced their construction of what became the literary canon and its conception of the Romantic era's aesthetic principles. I thus reveal how, especially through their portrayals of originality and sensibility, these male literary critics and authors continued and retroactively reinforced the historical and canonical erasure of scientific and women's literature through their contributions to establishing exclusionary, and often (to again use Coleridge's words) "contradictory" and "erroneous," standards or notions of Romantic-era aesthetics. As I explain, my goal in bringing to light these historical misrepresentations is to contribute to current movements within Romanticism, and literary studies more broadly, encouraging greater efforts toward inclusivity and accountability for authors as well as within scholarly understandings of this diverse era of literature.

PART I

Temporal Sensibilities
Circannual and Circadian Rhythms

1

Botany's Seasonal Disorder

Thomson's Progressive Time, Conjectural Histories, and the Backwardness of Spring

ONE OF the most intriguing mysteries in eighteenth-century botanical poetry is the origin and purpose of verses penned anonymously in a copy of the Lichfield Botanical Society's published translation of Carl Linnaeus's *System of Vegetables* (1783), titled "The Backwardness of Spring Accounted For, 1772." The poem is handwritten in the endpapers of a British Library copy of the translation, and its composition generally dated to the 1780s.[1] These verses humorously attribute continued wintry weather in May to the disorder of the plant kingdom, conveying blatant sociopolitical commentary regarding class hierarchies and national order, which supposedly will be remedied or restored through the English translation of Linnaean botanical classification. This chapter sheds new light on the poem not only by revealing its context within a broader subgenre of "Backwardness of Spring" poems but also by arguing that this subgenre responded to James Thomson's georgic *The Seasons*, which prominently engaged with botanical studies and contributed to establishing British conceptions of climate, time, sensibility, and conjectural history.

Thomson's *Seasons* (1726–46), the most popular English georgic of the eighteenth century and "the most fundamental English georgic of all time," celebrates British climate, vegetation, and geography and contrasts these elements with various other nations, particularly those existing in extremely hot or cold environments.[2] In doing so, Thomson also arguably contributes to the cultural movement of sensibility by reacting against earlier texts such as Thomas Hobbes's *Leviathan* (1651) and the expanded 1723 edition of Bernard Mandeville's *Fable of the Bees*, which interrogate English society to suggest that human actions are motivated by self-interest.³ Nationalistically reclaiming Britain as the ideal location for sensibility, especially in the sense of being able to feel and express sentiments of sympathy, love, joy, or, in some instances, even to feel at all, Thomson instead projects less desirable characteristics, including those connected with self-interest and *in*sensibility or a lack of feeling onto other nations. He anticipates numerous mid- and late eighteenth-century texts by naturalists and philosophers, such as Georges-Louis Leclerc, comte de Buffon, and Charles-Louis de Secondat, baron de La Brède et de Montesquieu, by associating nations' climates with national character, especially as shaping humans' physiological, emotional, moral, intellectual, and industrial capacities.

Within this ideology of geographical valuation and environmental determinism, Thomson Eurocentrically presents Britain's temperate climate as key to its sensibility as well as to its civil, moral, commercial, and intellectual achievements. This nation's climate thereby produces, in his language, "progressive" time, connecting continual cultural advancement with the progressing seasons that form a paradoxically cyclical, yet linear, temporality of forward movement.⁴ Thus, Thomson's conception of temporality moreover aligns with what some scholars call "imperial time" or "historical time," critiquing the inadequacies of such formulations for providing meaningful or accurate understandings of societies that do not reflect European temporal frameworks.⁵ Measuring other societies against British cultural standards, Thomson presents the climates of the "torrid" and "frigid" zones as functioning in a kind of atemporality because, within his poetic depiction, they are confined to the seasons of summer and winter, respectively.

Nevertheless, in the midst of Thomson's patriotic panegyrics, he also briefly provides hints for perceiving Britain in decline—in danger of slowed or halted progress, or even of regression. Significantly, in *Spring*, he envisions this potential for national degeneracy through the phrase "backward time," contemplating history as a means of both striving to recognize, and find remedies for, modern decay in British arts and morality (l. 927). In the context of Thomson's *Seasons*, this potential slide backward in Britain's historical stages of society also threatens to move the nation into temporal, moral, and intellectual states of torpor or insensibility and degeneration. In this way, as I show, Thomson participates in the construction of what is known as "conjectural history," or stadial theory, which had been debated since classical times and would become most connected with the Scottish Enlightenment, imagining human progression toward civilization through stages including hunting, pasturage, agriculture, and commerce. In such conceptions, Europe's supposedly "civilized" societies contrast with other populations that Thomson associates with the torrid and frigid zones and historical stages of advancement. Moreover, depictions of botany become bound up with these European standards of national hierarchy and conjectural history, especially through ideas relating to labor, commerce, sensibility, and the natural sciences.

Thus, I argue that, by predicating Britain's preeminence on its temperate climate and linear or "imperial time," Thomson's poem also paradoxically destabilizes this "progressive" conception of his nation, which thereby becomes vulnerable through challenges or disruptions to its forward seasonal movement. Thomson's notion of Britain's intertwining temporal, environmental, and cultural potential to slide backward on a climatic scale of stadial theory influenced some subsequent poets, including the author of the anonymous manuscript poem (which turns out to be not so anonymous), Richard West, Anna Barbauld, John Penn, and John Clare, to compose within the subgenre of Backwardness of Spring poems.[6] These later authors express varying levels of humor and concern when experiencing instances of seeming seasonal stagnation, especially in which the nation appears stymied in winter, signifying, as it did for Thomson, a potential lack of valued qualities, such as sensibility and industry. Therefore, within Backwardness of Spring poems, this perceived slowing of seasonal time that affects vegetal growth provokes fear also, through conjectural history, of Britain's regression, precariously linking the state of the nation with its changing environment.

Thomson's British Botany and Constructions of "Progressive" Time

In the tradition of georgic poetry, Thomson's *Seasons* addresses themes such as science, agriculture, and nation-building and would strongly influence subsequent descriptive, scientific poetry over the next century. While writing early portions of the poem, Thomson taught at Watts's Academy, an institute for popular study of Newtonian science; he dedicated *Winter* (1726) to Newton, and *The Seasons* itself is "famous for its celebration of Newtonianism," as well as of Whiggish politics.[7] In an essay specifically devoted to analyzing *The Seasons*, John Aikin, the late eighteenth-century critic and brother of Anna Barbauld, calls it a "rare" poem of "masterly execution" and "novelty of design," deserving "admiration" and "mature enquiry" because it "forms [a new] era in the art [of poetry] itself."[8] He states, "Thomson's *Seasons* is the original whence our modern descriptive poets have derived that more elegant and correct style of painting natural objects which distinguishes them from their immediate predecessors.... None of [the modern poets], however, have yet equaled [Thomson]" (v). In his *Essay on the Application of Natural History to Poetry* (1777), Aikin urges modern poets to achieve novelty by engaging with natural history and following Thomson's example of descriptive verse, for "it is in that truly excellent and original poem, Thomson's *Seasons*, that we are to look for the greatest variety of genuine observations in natural history."[9] The British naturalist Thomas Pennant likewise quotes from *The Seasons* in his *Arctic Zoology*, reverently calling Thomson "that just observer of nature, the naturalist's poet."[10] The georgic form, especially through Thomson's influence, arguably transmuted or disseminated into the movement of scientific literature that became prevalent in the late eighteenth and early nineteenth centuries among writers such as Charlotte Smith, Anna Seward, and Erasmus Darwin.[11] As Carson Bergstrom writes, "the 'new way' for poetry seems clear to Thomson ... a poet must adopt the mental disciplines demanded by the scientific method. A new poetic ethos might then arise to counter the 'Torrent of Barbarism' and bring about a 'Revival of Poetry.'"[12] Indeed, Thomson declares in *The Seasons*, "tutored by ['science'] ... Poetry exalts / Her voice to ages," thus championing scientific knowledge as imbuing verse with both novelty and lasting significance (*Summer*, ll. 1753–54).

Focusing on the seasons, Thomson's scientific, descriptive verse highlights the dynamics of the climate and geographical advantages enjoyed by Great Britain. He thereby helps establish what became the pervasive British belief that their climate was "an example of God's providential goodness to the island's people."[13] Although Thomson asserts Britain's "superior[ity]" and "own[ership]" over the "world" as the benevolent caretaker of "naked nations" that benefit from Britain's "exuberant," "rich soil" and "exhaustless granary" (*Spring*, ll. 72–77), as Suvir Kaul writes, "we should [also] note the cultural and moral arrogance embedded in" these statements, recognizing that Thomson is "best understood as a poet of empire."[14] Britain had long been a producer and exporter of wool and grain, establishing its imperial status within trade and commerce while concurrently importing the productions of other climates and countries. "Britannia," Thomson declares, now "commands / The exalted stores of every brighter clime, / The treasures of the sun without his rage" (*Summer*, ll. 424–26). Thus, through maritime navigation Britain imports useful productions of harsher climates to its own milder temperatures. This resonates with Joseph Addison's earlier pride in London's Royal Exchange as an "emporium for the whole earth."[15]

Born and raised in Scotland, Thomson was "perhaps the first important poet to write with a British, as distinct from a Scottish or English, outlook."[16] His poem attributes Britain's modern success to inward peace and unification, subtly praising the union of England and Scotland in 1707 as creating the nation's current strength and sensibility. In this vein, he looks to British history to admit that it did not always embody such a prosperous and civilized condition. He acknowledges that Britain once existed in "barbarous times," "[em]broil[ed]" in internal conflict because "disunited," but now embodies an "indissoluble state" that supports the growth of "wealth," "commerce," industry, "liberty," and civil government (*Spring*, ll. 842–48). Indeed, in *Summer*, Thomson attributes the prosperity of "Happy Britannia!" to its "Arts" and "Liberty," as well as its geographical and climatological situation, declaring, "Rich is thy soil, and merciful thy clime," praising the nation's natural and industrial productions (*Summer*, ll. 1442, 1443, 1446, 1448). He also patriotically portrays the British national character as imbued with a wisely moderate sensibility through particular "virtue[s]," describing Britons as "sincere, plain-hearted, hospitable, kind," but also, "when provoked, / The dread of tyrants" (ll. 1474–77). In addition to recounting some of the island's historical kings, heroes, and explorers, Thomson highlights Britain's intellectual and imaginative achievements, diminishing distinctions between the sciences and arts by moving seamlessly through references to Bacon, Boyle, Locke, Newton, Shakespeare, Milton, Spenser, and Chaucer (ll. 1479–1579). He thus invests poets with an agency and power related and equivalent to that available through the sciences in shaping the nation.

Within this context of British scientific achievements, Thomson repeatedly exemplifies British discoveries in botany as signaling the nation's interdependence of climate, geography, sensitivity, and industriousness, and as exhibiting "progressive" time, especially in his poem *Spring*.[17] Indeed, as Aikin explains, plants are etymologically bound up with the spring in the English language. as this "season derives its appellation" from "that universal *springing* of the vegetable tribes" at this time of year.[18] Emphasizing temporality within British nature, in *Spring* Thomson elucidates a calendar of flora, describing the chronological order in which particular flowers bloom over the course of the season: "The snow-drop and the crocus first, / The daisy, primrose, violet darkly blue, / And polyanthus of unnumbered dyes. . . . Then comes the tulip-race," and so on (ll. 530–32, 539).[19] Following this list of "endless" flowers by their common names, Thomson celebrates their "infinite numbers, delicacies, and smells" and displays knowledge of various botanical theories popular in the first half of the eighteenth century (ll. 555, 553). He attributes the intricacies of plant life to God's craftsmanship, reinforcing the Enlightenment justification for scientific investigations and discoveries as a means to better understanding God and His works. Thomson interweaves scientific references and terminology within his verse descriptions, enthusing:

> By thee [God] the various vegetative tribes,
> Wrapt in a filmy net and clad with leaves,
> Draw the live ether and imbibe the dew.
> By thee disposed into congenial soils,
> Stands each attractive plant, and sucks, and swells
> The juicy tide, and twining mass of tubes.
> At thy command the vernal sun awakes
> The torpid sap, detruded to the root
> By wintry winds, that now in fluent dance
> And lively fermentation mounting spreads
> All this innumerous-colored scene of things. (ll. 561–71)

Thomson portrays the plants as functioning in accordance with a divine plan that also ideally disperses them "into congenial soils" suited to their particular species, allowing them to revive in spring's warmer temperatures. Revealing the inner workings of botanical organisms, he here versifies the findings of Stephen Hales's *Vegetable Staticks* (1727), "the first full scientific account of the flow of sap in plants and the function of the leaves in plant respiration and nutrition."[20] In this scientific framework, Thomson describes the plant's system of sap vessels as the "filmy net" and "twining mass of tubes," and echoes "imbibe" from Hales's text as "the effect of the sun's heat in bringing moisture to the roots and raising sap by transpiration."[21] Thomson likewise employs "live ether" to refer to Hales's airborne "activating principle" that supports animal and vegetable life, while "fermentation" is Hales's word for its functioning.[22]

Further demonstrating his botanical knowledge, Thomson depicts contemporary debates over color theories through vegetation, gesturing toward Newton's spectrum. He poetically endorses the significance of

Various hues; but chiefly thee, gay green!
Thou smiling Nature's universal robe!
United light and shade! where the sight dwells
With growing strength and ever-new delight. (*Spring*, ll. 83–86)

Thomson here follows other naturalists in espousing "an argument for Providence," explained in Addison's *Spectator* no. 387, by which "the whole earth is covered with green, rather than with any other color, as being such a right mixture of light and shade, that it comforts and strengthens the eye instead of weakening or grieving it."[23] Through these botanical theories, Thomson's depictions of green organisms in *Spring* not only reveal the benefits of vegetation for humanity, particularly for the emotions and senses (which were key to scientific or empirical conceptions of sensibility), but also create consciousness of temporal advancement through vegetation's growth and mutability. And, in conjunction with this temporal movement, Thomson describes the British seasons as "roll[ing]" forward in a "ceaseless" motion (l. 1167).

Summer and Winter; or, Temporal Stasis

As noted earlier, in contrast to Britain's paradoxically cyclical and linear temporal movement through the seasons, Thomson depicts the torrid and frigid zones as less dynamic. Portraying this temporal difference in *Summer*, he emphasizes this season's seemingly slower progress even in Britain, explaining, "As the face of nature in this season is almost uniform, the progress of the poem is a description of a Summer's day" (37). Thomson's focus on summer's "uniform[ity]," confining his description of this season in Britain to a single day, thus limits the nation's usual sense of seasonal alteration merely to that of circadian rhythms; he then portrays this constrained climate as the perpetual state of nations in the torrid zone, where, according to him, it is always summer. In this zone he includes, for instance, Egypt and other African nations, India, Pacific and Caribbean islands, as well as Latin and South America.

In Thomson's view, this supposed suspension of seasonal movement has both positive and negative implications for those warmer nations.[24] Regardless of how hot the summer weather in Britain may seem, he assures it is more temperate than that of "the torrid zone," where they have "climes unrelenting! with whose rage compared, / [Britain's] blaze is feeble and . . . skies are cool" (*Summer*, ll. 632–34). In thinking about the relationship between weather and climate, as Jesse Oak Taylor explains, climate is "an abstraction" that cannot be directly experienced through the senses, while weather is immersive and "immediate"; nevertheless, "on some level we do experience climate every time we correlate the weather with broader patterns."[25] Portraying the torrid zone's climate as a conventional tropical paradise, Thomson depicts those nations near the "equator" as being temporally and vegetally "in eternal prime," with "unnumbered fruits," including the lemon, tamarind, pomegranate, and pineapple; yet this georgic convention, the "trope of bounty," also indicates the imperial ideology that plentiful resources available to these tropical societies enable a lack of industry (ll. 657–58, 664–85).[26] Moreover, depicted as breeding disease, and thus as unhealthy and stifling, this intemperate climate also creates "intemperate man" (ll. 1053, 1060). Displaying ethnocentric prejudices and repeating stereotypes likely gleaned from travel narratives, Thomson asserts that in these warmer countries the climate and the sun itself corrupt sensibility, breeding emotions associated with tyranny, jealousy, rage, and revenge, and suggests that these inhabitants are "unenlightened" due to being "cut off" from "society," as well as from "philosophy" or science (ll. 894, 884–90, 1758, 1730, 913, 940).[27] For him, Britain's success largely depends on maintaining its own national virtues while overcoming its isolation through "navigation bold, that fearless braves / The burning line [of the equator] or dares the wintry pole" through the "mad seas" and "rag[ing]" weather of these foreign regions in order to help develop "the rising world of trade" and "unbounded commerce" (ll. 1768–69, 1001, 985, 1006, 1012). In fact, it is arguably Thomson's interest in trade that causes him to value the torrid zone more highly than the frigid zone, despite his derogations. The warmer climates produce various fruits, spices, and other commodities that Britons desire, thus prompting those maritime mercantile voyages. Describing the torrid zone as stifling human industry but encouraging vegetable growth, Thomson portrays the frigid zone, on the other hand, as hindering both.

In his poem *Winter*, Thomson arguably directs both harsher criticism and greater hope toward the climate and culture of nations in the "wintry pole," including Scandinavian countries, Iceland, and Russia. Repeating from *Summer* a rhetorical technique of geographical comparison, Thomson contrasts Britain's milder winter with that of "the frigid zone, / Where for relentless months continual night ... reign[s]" in "solitary vast" and "unbounded wilds" (*Winter*, ll. 794–97, 804, 799). Although he portrays Laplanders of the frigid zone as being blessed and happy in their distance from certain aspects of European societies, such as "false desires" or luxuries and

FIGURE 1. "Winter," designed by William Kent, engraved by Nicolas Tardieu for the quarter edition of Thomson's *Seasons*, 1730. (Image courtesy of the Huntington Library, San Marino, California)

the "barbarous trade of war," he also describes this "simple" life as devoid of intellectual and commercial rewards (ll. 847, 844, 845). Then, imaginatively traveling into more frigidly bleak and desolate regions, Thomson describes these inhabitants as "the last of men," indicating their distant placement on his scale of conjectural history (l. 937). He writes,

> Here human nature wears its rudest form.
> Deep from the piercing season sunk in caves,
> Here by dull fires and with unjoyous cheer
> They waste the tedious gloom: immersed in furs
> Doze the gross race—nor sprightly jest, nor song,
> Nor tenderness they know, nor ought of life
> Beyond the kindred bears that stalk without. (ll. 940–46)

According to Thomson, this "gross race" of humans in their "rudest form" become so inactive and numbed by the cold that they possess no sensibility—no joy, wit, or feeling, and can only muster enough energy to function in a dehumanized capacity, on a par with animals (ll. 944, 949, 940). Pennant similarly denigrated the frigid zone, praising the Linnaean botanist Daniel Solander, following his 1772 voyage with Joseph Banks to Iceland and other northern regions for overcoming that climate: "The *arctic* Solander must remain a fine proof that no climate can prevent the seeds of knowledge from vegetating in the breast of innate ability."[28] Likewise, Thomson provides hope for redemption through "active government," which may "new-mould" this "neglected empire" whose cold climate has kept humanity in a state akin to that of "remotest time" (ll. 950–53). To prove this possibility, he exemplifies Peter the Great, the czar of Russia, crediting him with bringing elements of European advancement including mechanical tools, trade, arts, "civil wisdom," and "martial skill" home to his country, thus raising it from "sloth" and "ignorance and vice" to thriving prosperity and the "social" concerns of sensibility (*Winter*, ll. 969–75, 982).[29] Anticipating the theories of naturalists later in the eighteenth century, Thomson thus suggests the possibility of nations overcoming their climates' challenges and "new-mould[ing]" their futures through intense application of industry and intellect.

Conjectural Histories and "Backward Time"

Although the phrase "backwardness of spring," employed in many of the poems that I analyze in this chapter, refers to a delay in the season's arrival or the untimely continuance of wintry weather, Thomson's verses arguably indicate additional notions of temporal backwardness as well. *The Seasons* begins with an invocation, seeking to entice the emergence of temperate spring out of harsher winter, stating, "Come, gentle Spring, ethereal mildness, come" (*Spring*, l. 1). However, Thomson also applies the phrase "backward time" when analyzing Britain's "historic truth" to determine how "to raise her virtue and her arts revive" (ll. 927, 926, 929, 931).[30] He portrays the natural environment as inspiring a backward gaze at Britain's past as a way to understand its present shortcomings and thereby improve or facilitate its future advancements. Elsewhere in the poem, Thomson calls on the "sage historic Muse" to "conduct us through the deeps of time" and "show us how empire[s] grew, declined, and fell" as well as what makes nations continuously content and prosperous (*Winter*, ll. 587–91).[31] John Barrell points out the contradiction of *The Seasons* in presenting Britain as being superior, yet also in historical and modern need of intervention by its national heroes to ensure its progression and success.[32] Thomson's view into the past thus expands notions of backwardness and its applications in the context of his poem, particularly in his temporal, moral, emotional, and intellectual depictions of non-European nations.[33]

In this regard, I argue that Thomson contributed to evolving Eurocentric concepts of what became known as "conjectural history" or stadial theories speculating about the stages of development through which humanity progressed to reach its modern state. Conjectural histories became closely associated with the Scottish Enlightenment, with the most famous examples being by Adam Smith, Adam Ferguson, John Millar, and Henry Home, Lord Kames.[34] These thinkers posit that societies pass through stages of, first, hunting and gathering, then pastoralism and nomadism, agriculture, and, finally, a stage of commerce. Ferguson's *Essay on the History of Civil Society* (1767) proposes a three-stage theory of societies moving successively through savage, barbarian, and civilized periods and came to constitute "a nearly ubiquitous part of the implicit knowledge of the nineteenth century."[35] As the taxonomies of Linnaeus and other naturalists that impose hierarchies and classifications on the natural world became popular in the second half of the eighteenth century, so did concurrent and interlinked efforts to taxonomize human races and nations, reinforcing European prejudices and justifications for imperialism through "scientific" speculations.[36] In Ferguson's formulation, Europe and European colonies constitute "the civilized, liv[ing] in large, highly organized societies," representing "dynamic, enterprising commercial nations"; to non-European nations, he attributes "a lack of social change and progress," as well as "static social orders" and "unchanging, despotic" characters.[37] Although the stadial theories of Ferguson and other Scottish Enlightenment thinkers were published in the decades shortly following Thomson's *Seasons*, elements of this genre can be traced through earlier works of, for example, Hesiod, Ovid, Lucretius, Hobbes, and Locke, with Mandeville's *Fable of the Bees*, part 2 (1729), arguably the "first full-length conjectural history," published about fifteen years prior to the final, revised version of Thomson's poem, which, as I show, in many ways also represents stadial theory.[38] Whereas the works of Hobbes and Mandeville notoriously critiqued English society as hypocritically espousing ideals of virtue while acting self-interestedly, Thomson instead appropriates conjectural history to bolster Britain's alignment with traits of sensibility and industry.

Indeed, in *Autumn*, Thomson incorporates his own brief version of a conjectural history of humanity's advancement through stages of development, attributing this capacity for progression specifically to the characteristic of "Industry," which is accompanied by "labour ... sweat, and pain," and brings into being sensibility, "every gentle art / And all the soft civility of life" as the "raiser of human kind" (ll. 43–47). Without, or before the use of, industry, Thomson describes humanity in the state of a "sad barbarian roving mixed / With beasts of prey"—"a shivering wretch," "by Nature cast / Naked and helpless out amid the woods / And wilds to rude inclement elements" (ll. 69, 57–59, 47–49). In this earlier existence, humanity lacked "love," "joy," "peace," and "plenty," the friends and family of "home," and life was a "heavy, dark, and unenjoyed ... waste of time" (ll. 66, 65, 71–72). However, where humanity developed industry, there came into being not only the capacity for sensibility but also all of the arts, "mechanical powers," "wit," "elegance, and grace," "science, wisdom," commerce, and a free and just society that, according to Thomson, distinguishes civilization (ll. 77–118). Interestingly, in the midst of this conjectural history he repeatedly denounces the sport of hunting, which was generally considered integral to the initial stage of humanity in stadial theories, and instead attempts to inspire feeling or sympathy for the birds, rabbits, and deer that are thus "cruel[ly]" pursued and slaughtered (*Autumn*, l. 394; *Winter*, ll. 789–93). Thomson describes hunting as "barbarous" and "tyran[nical]" and encourages Britons to give up such endeavors (*Autumn*, ll. 384, 390; *Winter*, l. 949). Although Thomson maintains a strong overall tone of nationalism, he condemns what he views as manifestations of Britain's decline or lack of progress through its participation in slavery and hunting, as well as its shortcomings in various arts. In these regards, Thomson's thinking about interconnections of climate, sensibility, temporality, and the possible "backwardness" (as opposed to progress) of the nation and the natural world in the context of conjectural histories influenced subsequent poets with concerns about untimely winter weather, and its effects on vegetation and other aspects of their existence.

West, Barbauld, and Penn: Vegetation, Mythology, and "The Backwardness of Spring"

In the decades following the publication of Thomson's *Seasons*, there grew a subgenre of poems by authors including Anna Barbauld, Richard West, John Penn, and John Clare, among others, using variations of the title "On the Backwardness of Spring."[39] All of these writers were familiar with Thomson's poem, and some of them even specifically reference it in their verse. As I have shown, this poetic notion of the backwardness of spring echoes the opening of *The Seasons* in wishing for the emergence of spring out of winter. Moreover, these later poems' emphasis on the slowness or backwardness of the seasons denotes concern about this weather anomaly and lack of temporal progress that, through Thomson's poetic example, became associated with potential alterations in climate and thus in national advancement and success. For, when spring does not arrive as usual, it raises questions of why, and what this absence means. Jan Golinski recounts that "unusual atmospheric phenomena . . . had traditionally been interpreted as portents of dramatic political events or admonishments of a punitive God," and that, although over the course of the eighteenth century "hundreds of individuals began to compile daily weather journals" to discern the regularity of long-term patterns, unusual weather events still revived old superstitions: "Weather that departed from seasonal expectations posed obvious problems for agriculture and was taken as a sign that things were generally in disarray—a bad omen for human affairs."[40] Thus, it seems no coincidence that poems in this subgenre especially appear around the time when Britain began to struggle with, and ultimately lose, its American colonies. Nevertheless, these poems on the backwardness of spring also invoke fears of Britain's insular, national decline. Since Thomson fixed nations of the frigid zone in winter and aligned this temporal stasis with corresponding intellectual, moral, emotional, industrial, and various other kinds of stagnancies, later writers perceived the possibility of Britain losing its progressive changes of seasons as suggesting an altered trajectory or classification within the nation's own conjectural history.[41] In addressing the nation's experience and understanding of climate and temporality, some of these subsequent authors describe the backward spring in terms of mythology and natural observation, as well as its implications for sensibility and industrial agency.

Richard West's "Ode to Mr. Gray, on the Backwardness of Spring"—written in 1742, and thus after Thomson's original publication of *Spring* in 1728, but before his final revisions of that poem—remained unpublished until 1775.[42] In it, the poet seeks to "invocate the tardy May," displaying alarm at "sudden blasts" of foul weather that "drive the zephyrs from the skies" (ll. 6, 3, 4). Asking that spring instead "resume" the "balmy breath and flowery tread" of "accustom'd May," West emphasizes that this temporal and climatic departure is strange and unexpected for Britain (ll. 7, 9, 18). His speaker insists that "Nature" itself "upbraids" the spring's "stay" or backwardness and seeks to rouse May from this stupor or regression, calling for the month to "awake,""restore,""revive," and "arise" to aid nature's "labours" (ll. 13–16, 20). In West's depiction, without this seasonal progression, not only humans, but also trees, flowers, birds, and the weather, are all negatively affected. Thomson's poem portrayed the birds of non-European nations as "less melodious."[43] Likewise, in West's verses, the backward season threatens silent stagnation in Britain, for now "birds forget to love and sing" and "storms alone [in] the forests ring" (ll. 23–24). West references Pomona, the ancient Roman goddess of fruitful abundance, gardens, and orchards, known for protecting fruit trees and caring for their cultivation; yet her labors now "fade" as "a plaint is heard from every tree," and "each budding flow'ret calls for" spring (ll. 20–22). For West, only when spring progresses as expected can it bring the virtues of "peace, plenty, love, and harmony" to "every being" in "heaven and earth," gesturing toward this seasonal advancement as being in accordance with divine order (ll. 28–30). The present seasonal delay thus unsettlingly disrupts aspects of divine design, creating "[dis]harmony" and doubt, which the speaker struggles to quell through hope for an imminent return to Britain's "accustom'd" sense of temporal movement.

Drawing on other traditions within Roman mythology to address untimely weather, Anna Barbauld's "On the Backwardness of the Spring, 1771" (1773), responds to the season of a particular year. Recent editors of her poems note that the "Meteorological Diary of the Weather" in the *Gentleman's Magazine* "for March and April 1771 . . . shows many 'heavy, black, cold' days into the first week of April, and snow on 17 April."[44] The poem itself describes a snowstorm in May (l. 28). Barbauld includes a Latin epigraph from Virgil's *Georgics*, the source inspiring Thomson's chosen genre, encouraging seasonal movement by "chiding laggard summer and the loitering zephyrs." She begins each of her first two stanzas with the refrain "In vain," resonating with Thomas Gray's "Sonnet on the Death of Richard West" (wr. 1742, pub. 1775). Yet, unlike Gray's poem, Barbauld describes a sense of bereavement reflected in the natural world, portraying her femininely personified Spring as solitary, "mournful," "wounded," and covered in "tears" of "dreary" and "relentless sorrow" that bring "frequent showers" and "chilling dews" (ll. 13–16, 9, 10). Hearkening back to her Virgilian epigraph, Barbauld versifies Roman myth's explanation for how the seasons came into being, connecting her depiction of "mournful" Spring with Ceres, the Roman goddess of earth, agriculture, fertility, and maternal relationships. In this tradition, Ceres's daughter, Proserpina, was abducted by Pluto, the god of the underworld, and, unable to find her, Ceres stopped all vegetative growth, turning the earth into a desert through her grief. Jupiter intervened, ordering Pluto to free Proserpina, but Pluto first made her eat six pomegranate seeds, with the consequence that she must spend six months of the year with him in the underworld.[45] These months of Ceres's annual mourning over the separation from her daughter constitute winter, disrupting the previously continuous and perfectly temperate weather enjoyed by the world before Pluto's trespass.

Conjuring up this classical past, Barbauld's poem brings into view the eighteenth-century debate between the ancients and the moderns, so that, as Thomson also suggests through biblical references to Noah's flood, seasons become the result of a former "fall from grace," ensuring modern decline and failings, embodied in the natural environment (*Spring*, ll. 309–35). While the moderns were generally viewed as superseding the past in regard to the sciences and technology, the ancients were often lauded as maintaining superiority in the arts and their descriptions of natural phenomena, thus creating a feeling of modern artistic stagnation or failure which, in Barbauld's poem, manifests in this seasonal backwardness reflecting moderns' failings back to them. In Barbauld's depiction of spring 1771, time itself becomes "sad," slowly "mov[ing]" in "pensive hours" (l. 11). Indeed, echoing Thomson's depiction of the "frigid zone," Barbauld describes the backward Spring as "this frozen zone," which brings "no flowers" or birds' song, but only "clouds behind clouds in long succession" and "oppress[ive]," "heavy snows" so that this "waste fatigues the aching eyes, / And fancy droops beneath th'unvaried scene" (ll. 6, 8, 25, 21, 22–24). In this way, she portrays the experience of "this frozen zone" in terms of *in*sensibility, very like those that Thomson attributed to countries in the frigid zone. Britain's suspended wintry season produces a lack of mental, emotional, or poetic stimulation as imagination or "fancy droops" amid this intellectual "waste" and boredom of "th'unvaried scene," affecting both creative and temporal progression (ll. 23–24). She thus ends the poem with hopeful longing for the return of "genial sun-beams," "dissolving snows," and the "glad impulse" of May to separate or "loose" Britain from this "frozen zone" and restore its environment, as well as the individual's mind and emotions, to linear motion (ll. 25–28).

Barbauld's backward-spring poem is not the only work in which she expresses interest in and knowledge of the seasons' spatiotemporal progressions. Her *Lessons for Children* (1778–79) provides a "calendar of nature" in which she successively describes each month, and the natural events and human activities that can be expected to occur at those specific times.[46] Her work inspired her brother, John Aikin, to publish his own *Calendar of Nature* (1784) a few years later, which he dedicated to Barbauld, addressing a more advanced audience of young readers from ages ten to fourteen. He incorporates details of weather and natural history, especially from Benjamin Stillingfleet's Linnaean *Calendar of Flora* (1761), and frequent excerpts from Thomson's *Seasons* as well as from various other poets including his sister. For each month, Aikin relates the particular flowers and other vegetation that can be expected to appear. In his section about February, he quotes Barbauld's floral poem on the snowdrop:

> Already now the snowdrop dares appear,
> The first pale blossom of th'unripen'd year;
> As Flora's breath by some transforming power,
> Had chang'd an icicle into a flower.
> Its name and hue the scentless plant retains,
> And winter lingers in its icy veins.[47]

As I discuss later in this chapter, John Clare also associates transformative powers with spring and flora in his own poem on the season's backwardness. Still, the calendars of nature penned by Barbauld, Aikin, Stillingfleet, and others help to reinforce the sense of predictable temporal occurrences in weather and natural objects as described in Thomson's *Seasons*, creating feelings of dissonance and concern when those expectations fail to be met.

These emotions of disorientation also are expressed by the poet John Penn, whose "Ode on the Backwardness of Spring" (1797) joins specific

FIGURE 2. "The Snowdrop," from Robert John Thornton's *New Illustration of the Sexual System of Carolus von Linnaeus*, 1807; 1799–1810. The snow and icicles on the roof of the house in the background display the wintry season in which this flower blooms. (Image courtesy of the Linda Hall Library of Science, Engineering, and Technology)

reference to Thomson's nationalist ideologies with naturalists' theories about climate and human labor, prevalent in the second half of the eighteenth century.[48] Grandson of William Penn, the founder of Pennsylvania, John Penn was born in London and became the chief proprietor of the Province of Pennsylvania in 1775.[49] He lived in Philadelphia from 1783 to 1788 and returned to England in 1789, receiving compensation from Parliament for his land confiscated by the Pennsylvania legislature after the American Revolution. In addition to collections of poetry and a patriotic play about England's King Alfred—echoing Thomson's *Alfred: A Masque* (1740)—Penn also built a monument to Thomas Gray, whom he admired, and published a prose analysis of Virgil's "Fourth Eclogue" (1810).

Penn's "On the Backwardness of Spring" begins by interrogating Winter, asking why this season "still" persists in inflicting a "cheerless" "gloom" so late in the year (ll. 1, 3, 5). Through these disappointed hopes for spring, he sees a metaphor for human life, in which "thus oft our eager purpose fails" and "fruitless ends pursues," conjuring up feelings of failure and decline as well as unattained goals, perhaps resonating with his nation's and his own personal losses in the American colonies (ll. 11–12). Similar to Barbauld, Penn describes this wintry state as one of "mute Despondence" so that this is not the sensation of melancholy sometimes perceived interchangeably with wisdom (as in Hume's notion of "the disease of the learned") and as a potentially positive attribute of sensibility, but rather denotes inhibited mental and emotional activity and thus replicates the state Thomson attributes to inhabitants of the frigid zone (l. 17).[50] Interestingly, Penn's poem includes a note quoting lines from Thomson's *Spring* to substantiate the rejuvenation that is *supposed* to be occurring in Britain at this time of year. Gesturing toward a feeling of temporal dissonance, Penn draws on Thomson to paint a picture of the ordered progression of seasons and time, as if to say, "This is what Thomson promised would occur, so 'why' is it not here (l. 1)?"

However, in Penn's final stanza, he suggests that human industry ultimately will make the season advance, for "with the gay labours of the field, / The season hastes" (ll. 26–27). Here, human labor thus controls the weather and the progress of all aspects of culture and the natural world, restoring the absent "joys," "love," and "hope" (ll. 28–30). The ideas of naturalists such as Buffon asserting humans' ability to alter both themselves and the climate by industriously altering the environment had become, by the time Penn wrote in the late 1790s, a widely accepted commonplace. Oliver Goldsmith expressed Buffon's influence in this regard in "The Deserted Village" (1770), with the possibility of British colonists "redress[ing] the rigors of the inclement clime," and Barbauld later echoed this idea that "useful toil" could "New mould a climate and create the soil, / Subdue the rigor of the Northern Bear, / O'er polar climes shed aromatic air," in her poem *Eighteen Hundred and Eleven* (1812).[51] Such theories invested human agency with the capacity to control ecological factors of climate and soil that, within Enlightenment thought, shape national character. According to Buffon, a nation's "degree of civilization" depends on improvement of its land, so that, for instance, tangled and neglected vegetation denotes degradation of both nature and nation.[52] Indeed, for him, cultivation of the environment improves not only human inhabitants' civility but also their virtue and physical well-being, as well as their climate. As in Thomson's emphasis on progress through industry and sensibility, Penn places faith in human labor to again put the British nation and its spring on their forward-moving cycle.

Seward, Darwin, and Mundy: Humor, Climate, and Botany in a Backward Spring

A more humorous approach to the spring of the year following that addressed by Barbauld is taken by the poem mentioned in this chapter's introductory paragraphs, "The Backwardness of Spring Accounted For, 1772," penned in the endpapers of a copy of the second volume of the Lichfield Botanical Society's English translation from Latin of Linnaeus's *System of Vegetables* (1783).[53] Members of this botanical society consisted of Erasmus Darwin, Brooke Boothby, and William Jackson, a proctor in the Cathedral Court, and the manuscript poem jocularly promotes and justifies their translation of the Swedish naturalist's botanical system.[54] Although the poem appears anonymously, modern scholars generally attribute it to the poet Anna Seward, as she is referenced in one of the verses ("By the Muses who there on their *Seward* attended," l. 86), prompting Ann Shteir, for instance, to posit that "internal evidence points to Anna Seward" as the author; Sam George, Asia Haut, and others follow suit.[55] However, this internal reference also arguably makes it less probable that Seward is the author. It is important to be circumspect about such attributions, as Erasmus Darwin, for instance, sometimes assigned his own verses to Seward and at other times appropriated her poetry under his name, much to Seward's chagrin.[56] Further supporting the case for an alternate author, Seward's poetic friend William Hayley wrote verses similarly praising her:

> Ye gods, cried a bard, with a classical oath,
>   Who had order'd the bustos of Pope and of Prior;
> That on each side of Seward, who rivals them both,
>   They might properly honor that queen of the lyre.[57]

Numerous additional examples could be provided of Seward's poetic acquaintances admiring her in their verses, so her name's presence in "The Backwardness of Spring Accounted For, 1772," seems actually to discount, rather than support, her authorship.[58] Stylistically, the poem's "relaxed and jokey" tone appears more in line with Darwin's work and, in contrast, Seward's verses generally align with deeper emotional registers within the culture of sensibility.[59] Moreover, while Seward engaged with botany in various ways (as I explore in chapter 6), she also asserted that direct poetic treatment of the Linnaean sexual system was "not strictly proper for the female pen."[60]

Rather than Seward, the poem's author appears to be Francis Nöel Clarke Mundy, a wealthy Derbyshire landowner and magistrate, and a member of the Lichfield literary circle that included Seward, Darwin, Boothby, and Jackson.[61] Under a slightly different title, "The Backwardness of the Present Spring Accounted For, May 5th, 1782," this poem was posthumously published in 1830 along with several of Mundy's other works.[62] This published version of the poem includes a "Preface to the Reader," explanatory prose footnotes, and concluding acknowledgments, where Mundy writes that Darwin helped him with some of the botanical information, and "for this and other useful hints I desire he will be pleased to accept a very considerable share of the reputation arising from this work" (126). Mundy also admits that "the existence of this trifle [the poem] is entirely owing" to his efforts to win the praise of "the animated and elegant Miss Seward" and of Darwin and Boothby, "for nothing but my desire of shewing attention to their obliging and flattering notice of my productions could have prompted me to this undertaking" (126). Nevertheless, while Mundy asserts his authorship ("my productions"), this may be complicated by the fact that Darwin and Seward earlier anonymously contributed five stanzas to Mundy's pastoral poem *Needwood Forest* (1776).[63] Although the Lichfield literary circle sometimes collaborated by offering "useful hints" about how to improve one another's verses prior to publication, Mundy's acknowledgments seem to pinpoint Darwin as his main collaborator for the "Backwardness of Spring," and he wrote in hopes of earning the "notice" of Seward and the other Lichfield poets.[64]

In his poem, Mundy playfully reveals that it was the (interconnected) lack of botanical and social order in Britain prior to the Lichfield Botanical Society's translation of Linnaeus's system that caused spring's slowness this year. Although the anonymous manuscript version of the poem's title cites the spring of 1772, this is an error of transcription; the published poem shows that the year of the poem's backward spring was 1782, which makes more sense, as the society's translation was published in 1783.[65] Historical weather records note heavy snow in April 1782 in England's Midlands, including Derby and Lichfield, and in Scotland the season remained so "cold & backward" that the corn never ripened.⁶⁶ Perhaps the mix-up of the dates occurs because, George suggests, the poem's author was inspired by Linnaeus's speech "The Delights of Nature," delivered in 1772, describing botanical taxonomy as "homologous with a hierarchical model of human society and all its social classes and military ranks."⁶⁷ Since Linnaeus often drew human analogies to plants, his botanical system leant itself to comparisons with social order, and of course Darwin would famously exploit and explore these analogies in *The Loves of the Plants*, versifying Linnaeus's classifications of plants according to their reproductive parts. Darwin started writing that poem around 1779 and finished the first draft in 1784, continuing to expand and revise it until its publication in 1789, so Mundy may have been interacting with Darwin's manuscript verses as well.⁶⁸ Foreshadowing that later work's analogies between humans and plants, in Mundy's "Preface to the Reader" he writes that, according to Linnaeus, "the people of the vegetable kingdom consist of nine tribes: viz. Princes, Nobles, Highnesses, Patricians, Plebeians, new Colonists, Servants, Slaves, and Vagabonds," as well as "a distinction of sexes," so that "we have in plants Males, Females, Eunuchs, Mules, and Hermaphrodites" (III). Moreover, within "the marriages of plants . . . in some are found one husband with one wife, like the honest couples of our own species in this country," and in others are seen "one husband with a plurality of wives" and "concubines" (III). In "The Backwardness of Spring Accounted For," Mundy portrays the god Jupiter as "vexed to see things out of Season," for he had arranged a May Day procession to welcome Spring, featuring personifications of May, Venus, and Flora in the coach of Aurora (l. 15). However, in a statement of temporal stasis, linked with the threat of social and seasonal backwardness, Mundy declares, "Time seem'd to stand still" (l. 8). In his depiction, when May does appear, she looks out of character or anachronistic and is more in the guise of winter than spring because dressed "in strange dishabille . . . wrapp'd up in flannels and look'd very ill" (ll. 11–12).

According to Mundy's verses, Flora, the Roman goddess of flowers generally associated with spring, is the culprit holding up the procession (and progression) of May. When Jupiter demands an explanation, Flora praises "good order and laws" as necessary components of "civilization" to which "every nation" should aspire, yet complains that in Great Britain there is "an ignorant treatment of her and her System," causing "disorder" (ll. 38, 40, 39, 43, 45). Drawing a clear sociobiological analogy, in these years prior to the French Revolution, Flora laments that her botanical "government" is "strangely distorted" with a "leveling spirit" so that "rank and high titles . . . have no merit," there is "irregularity" and "want of distinction of Sexes & Classes," as well as "confusion of Manners and Morals" (ll. 48, 47, 50, 49, 66, 67). Moreover, Flora elucidates that her "empire" is in danger of being "dissever[ed]," and of her "*Colonists* claim[ing] independence for ever" (ll. 98–99). Since this poem responds to the weather of 1782 and thus follows Britain's loss of the American War, its lighthearted remarks about the prospect of losing colonies due to their claiming of independence may seem unpatriotic and less appropriate for publication, perhaps helping to explain why the poem remained unpublished for so long. Nevertheless, more broadly speaking, interconnections between sociopolitics, weather, and temporality find precedence in Thomson's poem as well.

To remedy these "disorder[ly]" circumstances, Flora entrusts Darwin and the other translators to be her "administration" and "great legislators," who come to the rescue by "prescrib[ing] / The laws, rules, and habits of every [vegetable] *Tribe*" and inspiring "England" to botanical study (ll. 105, 113–14, 119). Flora assures that once this translation of Linnaeus's work is published, then the ranks and orders among vegetation will assume the correct "decorum," and only at such time will she join the procession and allow spring to come forth (l. 165). Convinced by Flora's reasoning, Jupiter agrees to delay or "put back the Seasons," and thus Mundy comically makes the expected progression of Britain's weather, temporality, sensibility, intellectual development, and sociomoral structure dependent on the Lichfield Botanical Society's forthcoming translation (l. 168). Despite Mundy's humorous tone in portraying this 1782 weather anomaly, the poem's implications correlate with Thomson's association of climatic stasis and Britain's potential to slide backward on a scale of conjectural history, causing greater "disorder" and social, emotional, and moral "confusion."

Fascinatingly, given Darwin's contribution to Mundy's poem, and further connecting with Thomson's *Seasons*, stadial theory originally comprised the primary focus of Darwin's final long poem, *The Temple of Nature, or the Origin of Society* (1803). Drafts of that poem reveal that its initial title was "The Progress of Society," which then became "The Origin of Society," and it was "almost certainly" the publisher, Joseph Johnson, who adopted another alternative that Darwin had considered, *The Temple of Nature*, as its title after the poet's death in 1802.[69] Darwin's early drafts of the poem outline a plan to versify "the five ages of society," indicating those of "hunting, pasturage, agriculture, commerce and—lastly and triumphantly— philosophy (that is, science)."[70] In the published version of the poem, a footnote in the first canto espouses a formulation of conjectural history in which "some parts of the earth and its inhabitants appear younger than others," so that "the greater height of the mountains of America seems to show that continent to be less ancient than Europe, Asia, and Africa; as their summits have been less wasted away, and the wild animals of America, as the tigers and crocodiles, are said to be less perfect in respect to their size and strength; which would show them to be still in a *state of infancy*, or of *progressive improvement*."[71] By endorsing this conception of America as having the potential for "progressive improvement," he could be said to align with the ideas of Buffon, who also famously contributed to anxieties about the colder climates that had concerned Thomson.

According to Buffon, the earth initially formed in a molten state and thus was in a gradual process of cooling that would persist until becoming uninhabitable and dead.[72] However, Darwin proposes an ingenious solution to this theorized threat of increasingly cold global climates. In the footnotes of his *Economy of Vegetation* (1791), he writes that "there are many reasons to believe from the accounts of travelers and navigators, that the islands of ice [icebergs] in the higher northern latitudes as well as the Glaciers on the Alps continue perpetually to increase in bulk" and "the northern ice is the principle source of the coldness of our winters."[73] Darwin thus makes the radical suggestion that, rather than "destroying their sea-men and exhausting their wealth in unnecessary wars," European nations could "unite their labors to navigate these immense masses of ice into the more southern oceans," so that "the tropic countries would be much cooled by their solution, and our winters in this latitude would be rendered much milder for perhaps a century or two, till the masses of ice became again enormous."[74] Indeed, he speculates that Britain's unseasonably "cold winds and wet weather which sometimes happen in May and June" in England are due to icebergs "accidentally floating from the north."[75] Of course, in the context of Darwin's *Botanic Garden*, the most immediate benefit in eliminating winter's cold weather, and the accompanying "fiend of frost," lies in avoiding its harm to plants and agriculture.[76] Nevertheless, his proposal that European nations cease to wage war and instead work together to pursue the democratic and humanitarian ideals of achieving a temperate, universal climate appears to offer a scientific answer to the disparities of climate and national character that Thomson highlights in *The Seasons*. It is difficult to overstate the significance of Darwin's suggestion since, in Enlightenment determinism, propounded by various naturalists and perhaps most famously by Montesquieu, a nation's climate dictates not only body type and skin color but also virtually every aspect of inhabitants' identity, from temperament to sexual potency.[77] Thus, within this theorization, Darwin's hope of universalizing the climate also would eradicate national differences, potentially producing a globally homogenous nature and nation, an unvarying human race of worldwide equality.

Clare's "Spring" and Thomson's Legacy of Anxiety for British Climate and Progress

While staying at a village inn in 1798, Samuel Taylor Coleridge and William Hazlitt found a worn-out copy of Thomson's *Seasons* lying on a window seat, provoking Coleridge to exclaim, "*That* is true fame!"[78] A few years later, in the summer of 1806, a Helpston weaver showed a fragment of *The Seasons* to the then thirteen-year-old John Clare, who "never forgot the sensation—a 'twitter of joy' in the heart, he called it—that he felt upon reading the first few lines" of Thomson's *Spring*.[79] Clare most appreciated *The Seasons* for its descriptions of the weather, landscapes, and "Nature in all her varied moods and colors," and was intent on buying a copy; it would be the first book of poetry he possessed, and it inspired the first poem he ever wrote, "The Morning Walk."[80] When Clare later published his poem "The Backward Spring," in his volume *The Rural Muse* (1835), he displayed Thomson's continued influence while creating a memorable lyric of sensibility in its own right. Reading as a love poem to a personified Spring, Clare's verses echo Thomson's opening lines, wooing the season to "come" forward.[81] He conveys the weather's untimeliness, depicting the "too tardy Spring" in "Winter's apparel," but avers that it is now "no such thing," for "care and Winter are gone" (ll. 10–11, 12, 24). Repeatedly calling Spring "my charmer," Clare emphasizes this season's transformative powers, capable of conjuring "the flowers on the hawthorn, / Oak-balls on the tree" and "charm[ing] / Into summer the winter, / To sunshine the storm," as well as "turning pools into mirrors, / And silence to glee" with its "presence" (ll. 17–18, 42–44, 37–38, 47). However, the spring is not the only entity within the poem possessing the ability to "charm" or "change" the environment, as Clare aligns the poet and poetry with this altering potential as well: "Even snows, 'neath thy feet, / I could fancy to be / A carpet of daisies. / The rime on the tree / Would bloom in thy smiling" (ll. 25–29). Thus, the speaker's "fancy" or imagination acts as another kind of charm that can transmute one thing or season into another, if only in the mind, and Clare easily puns on "rime," or frost, as "rhyme," to imbue poetry with this transformative potency.

Finally, Clare ends his poem about untimely and atypical weather by suspending time and weather as well as his verses' theme of transmutability, suggesting instead the possibility of a perpetual spring. In the last lines of "The Backward Spring," he addresses the season, stating, "Though without thee I feel / What a desert would be, / I should think, in thy presence, / 'Twas Eden with me" (ll. 45–48). By conflating Spring with Eden, Clare arguably gestures toward Thomson's discussion of this theme. In *Spring*, Thomson references Thomas Burnet's *Sacred Theory of the Earth* (1684) which posits that, prior to Noah's flood, the Earth enjoyed a "perpetual Spring," but, following that universal deluge, the four seasons came into being: "There was no Diversity or Alternation of Seasons in the Year, as there is now."[82] Quoting Virgil, Burnet writes, "All the great World had then one constant Spring, / No cold East-winds, such as our Winters bring."[83] Whereas Thomson critiqued the national effects of climatic stasis in summer or winter, he ends *The Seasons* with a glance forward from winter to spring in a metaphorical sense of moving from the winter of life, and thus death, into "one unbounded Spring" as the state of Heaven (*Winter*, l. 1069). According to Burnet, ancient and Christian writers alike have "celebrated the perpetual Spring and Serenity of the Heavens in Paradise."[84] By suspending spring, Thomson Eurocentrically aligns that hopeful prospect of the afterlife with a continuously temperate climate indicating perpetual rejuvenation—botanical, emotional, and otherwise—and now positively portrays the concept of temporal and seasonal stasis.

In the mid-eighteenth century, Thomson's *Seasons* established for subsequent writers an understanding of Britain as developing within a mode of "progressive time" due to its particular climate and geographical or environmental conditions. In contrast to this British ideal, fostering an intellectual, moral, and industrious national character, Thomson described what he perceived as more static climates, temporalities, and characters in non-European nations. However, by plotting these various societies along a scale of conjectural history that moves toward a Eurocentric notion of commerce, sensibility, and civilization and making the British ideal contingent on its climate and temporality, Thomson additionally engendered concerns about national decline when those expected standards of British progress fail even briefly. As Thomson most denigrated nations of the frigid zone associated with winter, the unusually long continuation of that season in certain years inspired varying levels of humor, resistance, and anxiety in British writers and the population at large in later decades. This is evidenced not only in these remarkable "Backwardness of Spring" poems but also in, for example, reactions to the nation's strange, cooler weather resulting from volcanic eruptions elsewhere on the globe both in 1783, causing for Britain what has been called a "mini ice-age," and in 1815, resulting in the following year's famed "year without a summer," which devastated crops and brought famine.[85] In these instances, an underlying tension stoked by Thomson's legacy of associating climate, vegetation, sensibility, and conjectural history haunted those experiencing such weather anomalies with questions of whether the environment reflected a national decline, and the need for some measure of either human or divine intervention to set things back on course.

Continuing to think about vegetation, sensibility, and temporality, the next chapter moves from these concerns with botany's seasonal or circannual events to a more minute focus on certain flowers' daily, circadian rhythms during their times of blooming. One of Linnaeus's most fascinating botanical theories involved the configuration of "floral clocks." Inspired by this idea, several well-known British poets of this era differently formulated flowers' time-keeping abilities, contributing to scientific debates about mechanism and organicism, while displaying evolving relationships between botany, sensibility, and poetry.

2

Linnaeus's Botanical Clocks

CHRONOBIOLOGICAL MECHANISMS IN THE SCIENTIFIC POETRY OF ERASMUS DARWIN, CHARLOTTE SMITH, AND FELICIA HEMANS

WHEN JAMES COX, the renowned inventor and jeweler, opened his London museum in 1772, revealing to the public a series of elaborately crafted mechanical clocks and automata of exotic animals, plants, and human figures, four of his twenty-three "magnificent" and "useful" pieces showcased mechanical flowers "unfolding and closing again like nature."[1] In his final and "most distinguished display," bouquets of these unfolding flowers contained in the center flower of each "a curious timepiece," keeping the rhythms of the plants' movements.[2] Cox's mechanical flowers, associating nature, motion, and time, dramatize one side of contemporary debates about whether the movements of biological plants more closely analogized such passive clockwork automatons, or instead might illustrate a form of sentiency, inviting human analogy.

British Romantic–era poetry variously represents flowers along a temporal spectrum spanning from fleeting to time-transcending, engaging with these connected contentions about plants' physiological capacities for motion and feeling.[3] For example, Percy Shelley's *Sensitive-Plant* (1820) examines the species perhaps most central to the period's investigations into vegetable movement, the *Mimosa pudica*, a Brazilian plant that closes and recoils its leaves when touched. In 1729, the astronomer Jean Jacques d'Ortous de Mairan performed the first-known experiments in chronobiology, or the study of "adaptations evolved by living organisms to cope with regular geophysical cycles in their environment," displaying the fact that the spontaneous daily rise and nightly fall of the sensitive plant's leaves persist even while kept in the absence of light under constant conditions.[4] Earlier naturalists noticing other species' leaf movements described them as passive reactions to environmental stimuli, particularly the sun. De Mairan's experiments suggested that plants' physiological rhythms may be regulated not merely by environmental factors, but from within.[5] Over the course of the century, botanists improved on his work, generating new ideas about plants' timekeeping abilities and possession of internal clockwork, or what we now call circadian rhythms, directing plants' movements and prompting questions about the extent of their agency.[6] In Shelley's poem, vegetable species, and especially the *Mimosa pudica*, sensitively interact with their environment and respond to the death of their female caretaker, ultimately becoming emblems of ephemeral beauty themselves by succumbing to seasonal changes. I argue that such verse treatments of plants and flowers participate in a larger contemporary framework in which naturalists and poets explored and debated plants' movements, sentiency, and timekeeping mechanisms, particularly in relation to what became known as the "floral clock."

The popularity of botanical studies soared in Britain in the latter half of the eighteenth century, especially due to the Swedish naturalist Carl Linnaeus, whose work also helped inspire interest in flowers' movements and daily times of opening.[7] In his *Philosophia Botanica* (1751), Linnaeus, renowned for his system of botanical classification, which ordered plants according to their sexual parts, proposed groundbreaking investigations into horological processes in the vegetable world. He ambitiously called for "floral calendars" charting circannual developments of botanical species to be "completed every year in every province," and "floral clocks" functioning "under any climate" "to be worked out according to the watches of the plants, so that anyone can make calculation of the hour of the day without a clock or sunshine" (297). Advocating the *Horologium Florae* or floral clock, he refers to the phenomenon in which some flowers "watch," or open and close, at specific hours of the day and night, prompting him to theorize that one could arrange certain flowers into the face of a clock and know the time simply by perceiving which flowers are open at a particular moment.[8]

Linnaeus, in recruiting these botanical observations from naturalists around the globe, brought attention to plants' horological precision, which also appealed to many contemporary poets. For instance, James Grainger's georgic *The Sugar-Cane* (1764), which is rich with notes about West Indian natural history and Linnaean botany, follows the

FIGURE 3. A depiction of the flower clock designed by Linnaeus. The left half of the figure (6 a.m.–12 p.m.) displays different species' times of opening; the right half (12 p.m.–6 p.m.) shows times of petal closing (except evening primrose, which starts to open its flowers after 5 p.m.). Note that some species can act to time both morning and afternoon events. (Blumenuhr, drawing: Ursula Schleicher-Benz, after Carl von Linné, © 2022 Jan Thorbecke Verlag, Verlagsgruppe Patmos in der Schwabenverlag AG, Ostfildern/Germany www.thorbecke.de)

naturalist's injunction to record floral clocks from different geographical regions. Contributing knowledge of Caribbean plants' timekeeping abilities, Grainger explains that the "broom-bush . . . may, with propriety, be termed an American clock; for it begins every forenoon at eleven to open its yellow flowers, which about one are fully expanded, and at two closed. The jalap, or marvel of Peru, unfolds its petals between five and six in the evening, which shut again as soon as night comes on."[9] Since variations in climate and biogeography affect the timing of flowers' blooming, British poets and naturalists sought to configure a *Horologium Florae* specific to their nation.

In fact, Erasmus Darwin, Charlotte Smith, and Felicia Hemans each versified the botanical clock, producing different depictions and ideological agendas for this Linnaean device. All three poets contribute to scientific controversies about, for instance, whether plants' movements and potential sentiency originates in systems within these organisms, or in passive reactions to external stimuli, with ramifications for materialist philosophy's assertions that biological species function comparably to machines. The Linnaean analogy of a clock or watch lends itself to understanding plants' timekeeping operations as involuntary and mechanistic, yet Linnaeus's analogies elsewhere between humans and green organisms (in which, for example, parts of plants represent males and females capable of "marriage") suggest possibilities for imputing reasoning agency. Thus, while Darwin poetically portrays the floral clock primarily in mechanical and sexual terms, Smith explores these horological flowers in relation to sensibility and organic agency, and Hemans displays the era's later shifts away from both Linnaean botany and scientific poetry and instead toward more introspective concerns with human feeling and experience. These alterations in poetic depictions of the floral clock thereby signal some of the broader changes that occurred during that period with the increasing, separate professionalizations of literature and the sciences, especially by the 1820s and 1830s.

Darwin's Watch: Sexuality and Mechanical Nature

Erasmus Darwin versifies the Linnaean system of botany in *The Loves of the Plants* (1789), the second part of his long scientific poem *The Botanic Garden*, capitalizing on this system's concern with sexuality that excited frequent comparison between plants and humans while generating a potential for sensationalism that generally attracted rather than deterred interest.[10] Linnaeus's plant classifications simplified the botanical sciences and invited amateur participation by men and women alike. His systematic approach and "invention" of binomial nomenclature provided an efficient and "useful technology" so that, as Lisbet Koerner states, "his lasting contribution to knowledge . . . was his patient labor to mechanize and standardize the science of botany."[11] Among naturalists throughout the eighteenth century, mechanical and vitalist philosophies competed for prominence in physiological ideas. As Susannah Gibson states, "Plants were most likely to be either called 'Newtonian' and so described as hydraulic systems that followed mechanical laws; or they were living, feeling, perceptive beings that were capable of a certain degree of voluntary action."[12] The natural philosopher Stephen Hales formulated the concept of a "Newtonian vegetable" in his *Vegetable Staticks* (1727), describing plants as hydraulic machines that could be explained in numerical terms, especially the weighing and measuring of plant fluids (156). In contrast, Thomas Percival, a Manchester physician who knew Darwin, claimed in his vitalist article "Speculations on the Perceptive power of Vegetables" (1785) "not only that plants had a life force, were capable of spontaneous motion, and experienced sensations, but also that they had genuine powers of perceptivity" (160).[13] Amid such contentions, and despite using personifying language in his writing about plants, Linnaeus was often identified with mechanism, and one of his greatest rivals, the French naturalist Georges-Louis Leclerc, comte de Buffon disparaged his taxonomies as mechanistically abstract and dependent on observations of surface phenomena.[14]

Similarly, as recent scholars note, attacks on Darwin's poetry by contemporaries largely focused on its perceived mechanism. According to Packham, while Wordsworth's preface to *Lyrical Ballads* espouses an "uncompromisingly organic and vital poetry," he degrades Darwin's floral personifications as a "mechanical device of style."[15] Southey also categorizes Darwin's verse as "poetical machinery" that appeals to "materialists of fine literature" because of its art and polished style, but lacks the "life and feeling of poetry."[16] And Anna Barbauld coldly praises Darwin's *Botanic Garden*, stating, "His verse is a piece of mechanism as complete in its kind as that which he describes," characterizing his poetry as an "artificial species of excellence."[17] Modern literary critics have responded both by justifying these accusations, for instance, through exploring Darwin's "materialist philosophy of mind" and poetry as indebted to Lucretius's epicureanism, and by attempting to redeem his work as instead representing "vitalism" and "concern with organic life," standing in "opposition to a natural philosophy founded on mechanical models."[18] This critical confusion results from Darwin's negotiations of mechanism and vitalism in his poetry and prose, exemplifying what Peter Hans Reill terms "Enlightenment vitalism," comprising "the imperative to mediate between extremes in which harmony functioned as its overriding metaphor. Its creators consciously sought to retain elements of mechanism and animism (the closest eighteenth-century equivalent of organicism) but placed them in a new context."[19]

The floral clock episode in *The Loves of the Plants* represents one such "new context" in which Darwin struggles to harmonize mechanism and animism, emphasizing Linnaeus's mechanistic analogy in the "Watch of Flora." Although scholars often note Darwin's use of analogy in this scientific poem that likens plants and humans, he was also an inventor of various machines, and his poem displays these overlapping interests in technology and the natural world.[20] Anticipating Reill's notion of Enlightenment vitalism, Maureen McNeil argues that Darwin "was interested in defining distinctive organic functions and movements, and, thus can be regarded as a 'vitalist'"; however, "his theory was firmly within the materialist tradition associated with Hartley and Priestley [with whom Darwin was friends], and . . . his physiological theory was formulated around the model of the body as machine," displaying the coexistence of these different categories within his scientific thought.[21] In one of his many lengthy scientific footnotes, Darwin lists the twenty-one timekeeping flower species most prevalent to England from Linnaeus's "watches" of Flora, additionally providing their scientific and common names as well as the times of their opening and closing with the caveat that "as these observations were probably made in the botanic gardens at Upsal [Sweden], they must require further attention to suit them to our climate" (63).[22]

Subtly participating in disputes about plants' abilities to act and react in relation to their surroundings, Darwin personifies three of these twenty-one flowers in his initial verses about the floral clock:

> The gentle LAPSANA, NYMPHÆA fair,
> And bright CALENDULA with golden hair,
> Watch with nice eye the Earth's diurnal way,
> Marking her solar and sidereal day,
> Her slow nutation, and her varying clime,
> And trace with mimic art the march of Time;
> Round his light foot a magic chain they fling,
> And count the quick vibrations of his wing. (ll. 163–70)

Here, Darwin's flowers "trace" the movements of the "Earth," of their environment, in "mimic art," referencing a passively responsive form of agency that embodies an answer to contemporary debates about the degrees of, or lack of, agency (e.g., capacities for volition, sensation) in what many characterized as nonsentient matter, including plants. In chronobiological terminology, Darwin's "mimic art" resembles modern understandings of how plants' internal or endogenous rhythms display photoperiodism, meaning responses to seasonal changes in day length, "by *entraining*, or locking on to, the driving oscillation of the environment in what is called *photic entrainment* or *photoentrainment*."[23] Physically sensing solar, external, or exogenous rhythms in their environment, these flowers "count the quick vibrations of [Time's] wing." In this way, Darwin presses the chronobiological importance of both internal and external rhythms in producing organisms' timekeeping capacities.[24]

"Marking" or "watch[ing]" environmental signals of time's daily progress, Darwin's plants "trace" the motions of the sun and stars in a cosmological mechanism known as "nutation," possessing astronomical, botanical, and mechanical connotations. Drawing on nutation in astronomy that measures the rocking or swaying motion in the earth's axis of rotation, caused by forces including the sun and moon, Darwin transfers this astronomical movement into the movements of these three flowers as they physically "watch" and follow, or turn with, the progress of these celestial bodies across the sky.[25] Darwin's anthropomorphized flowers thereby enact botanical nutation, the slight curving or circular "motion" in which they quit "their perpendicular direction, present their surface directly to [the sun], and follow its situation in its diurnal course."[26]

Associating botany with astronomy, Darwin references Linnaeus's radical assertions that rank new progress in botany as superior to Newtonian astronomical discoveries, claiming that botanists "will provide something of much greater use to the public."[27] Linnaeus's call for floral calendars and floral clocks thus envisions a means to revolutionize human thought about time in a way that, according to him, may surpass even Ptolemaic and Keplerian astronomy or Newtonian mechanics.[28] His floral clock revives earlier correlations between botanical and astronomical means of telling time, as when Pliny the Elder personified Nature, stating, "I have given you plants that mark the hours.... Why then do you still look higher and scan the heavens themselves? Lo! you have Pleiads at your very feet."[29] When in possession of botanical clocks, astronomical timekeeping becomes superfluous. Darwin's use of "nutation" depicts these plants' movements as not only tied to astronomy but also as mechanical, since the term additionally refers to any intended behavior of a mechanism.

Portraying flower species as representing the hours of their opening on the face of the Watch of Flora, Darwin depicts their timekeeping mechanisms as machinery:

> First in its brazen cell reluctant roll'd
> Bends the dark spring in many a steely fold;
> On spiral brass is stretch'd the wiry thong,
> Tooth urges tooth, and wheel drives wheel along;
> In diamond-eyes the polish'd axles flow,
> Smooth slides the hand, the balance pants below.
> Round the white circlet in relievo bold
> A Serpent twines his scaly length in gold;
> And brightly pencil'd on the enamel'd sphere
> Live the fair trophies of the passing year. (ll. 171–80)

Configuring the floral clock conceit in its most extreme form, Darwin analogizes nature to machinery so that biological organisms become metallic, systematic, and predictable parts of this mechanical watch of nature. According to Descartes's materialist theory of the "beast-machine" in his *Discourse on Method* (1637), which influenced many Enlightenment theorists, the "laws of mechanics" are identical with the laws of nature.[30] His conception of dualism set apart the human mind from our physiological structures, as well as from animals and plants, which he likened to inanimate machines, implying, for instance, that the operation of a human arm is "like" that of the "arm" of a machine constructed from levers and pulleys.[31] Adopting mechanistic analogy, Darwin's anthropomorphized flowers retain metaphorical body parts (e.g., "tooth," "eyes"), functioning as automated, regular movements of springs and interlocking cogs that force forward the "smooth slid[ing] . . . hand," pointing to flowers in place of numbers on the watch face to reveal the time. This mechanized depiction undermines the flowers' previous appearance of even imitative agency, now passively operating as small parts of a larger, regulated process. They embody temporality both through their telling of time within the floral clock and as "the fair trophies of the passing year." This image of ephemerality contained within a symbol of eternity, the circle, the "serpent" with its tail in its mouth, presents time as at once both fleeting and never-ending. Darwin's flowers thus prove the bromide of mortality, "the soft bloom of Beauty's vernal charms / Fades in our eyes, and withers in our arms," preparing us to think of that which is of more lasting significance (ll. 195–96). However, where another writer might then turn his thoughts to the divine, Darwin portrays the destruction of religion.

Although eighteenth-century naturalists often reconciled science and religion in expressions of natural theology or physico-theology, validating mechanistic views of nature with such well-known metaphors as the deist formulation of God as watchmaker, setting or keeping the universe in motion, Darwin rejects this notion.[32] For him, these horological mechanisms of botany and astronomy participate in a cosmology of reason, not religion, and Time becomes personified as the force through which new knowledge and biological organisms come into being:

> Here *Time*'s huge fingers grasp his giant-mace,
> And dash proud Superstition from her base,
> Rend her strong towers and gorgeous fanes, and shed
> The crumbling fragments round her guilty head.
> There the gay *Hours*, whom wreaths of roses deck,
> Lead their young trains amid the cumberous wreck;
> And, slowly purpling o'er the mighty waste,
> Plant the fair growths of Science and of Taste.
> While each light *Moment*, as it dances bye
> With feathery foot and pleasure-twinkling eye,
> Feeds from its baby-hand, with many a kiss,
> The callow nestlings of domestic Bliss. (ll. 181–92)

Figured in colossal form, the concept of Time in nature, long used as evidence of God's craftsmanship, now smashes such religious associations, replacing them with human understanding of observable (and, here, self-reflexively temporal) phenomena like the floral clock.[33] Instead of employing the clockmaker theory, Darwin "plant[s] the fair growths of Science and of Taste," stressing the incompatibility of knowledge and superstition in the interest of scientific and cultural advancement (l. 188).

Demonstrating physiological progress, Darwin's floral clock episode culminates in sexual reproduction, represented in Time's smallest and most fleeting form, the "light *Moment*," the clock's "baby-hand," ticking off "pleasure-twinkling" seconds of "domestic Bliss" (ll. 189, 190, 191, 192). In religion's absence, the production of offspring becomes our only immortality.[34] Throughout Darwin's poetic works, and especially in *The Temple of Nature* (1802–3), sexual reproduction is the mechanism of progress and variation in species, producing new and perfectible organisms, as well as emphasizing the importance of change over time.[35] In chronobiological studies, reproduction, of course, represents the most likely explanation for the existence of flowers' timekeeping mechanisms, displaying "intricate pollination symbioses with insects that are dependent on biological timing," as many plant species depend on bees for cross-fertilization and thus evolved timing strategies, for example, to keep their pollen dry or to ensure insects deliver the pollen of one flower to another of the same species.[36] Likewise, for Darwin, horological flowers exemplify reproductive efficiency inherent to biological progress in both individuals and species.

In Darwin's floral clock scene, the tensions between mechanistic fixity and organic change over time, as well as between conceptions of exogenous and endogenous stimuli motivating plant movements, limit these organisms' potential for sentiency and agency.[37] Linnaeus himself stated, "Vegetables have life, but no sensations."[38] However, elsewhere in *The Loves of the Plants* (e.g., in his descriptions of the Alcea and Iris), and in other works, Darwin more forcefully challenges the materialist distinction of plant and animal automatism from humanity's capacity for thought, sentience, speech, and skill development. Although some eighteenth-century naturalists attributed such qualities to animals, Darwin additionally and more radically argued for sentiency in plants, especially in his final prose scientific text, *Phytologia; or The Philosophy of Agriculture and Gardening* (1800). Espousing early notions of evolution, he calls plants "an inferior order of animals," endowing them with "sensibility" and "volition," demonstrated through their various movements that, according to him, prove plants not only capable of "associations of motion, or habits of action" but also as thus in possession of "a brain or common sensorium belonging to each bud," which "supplies the spirit of vegetation since it exists in all buds in their most early state ... and evinces their individuality."[39]

Moreover, by 1800, Darwin deemed Linnaeus's theory of vegetable reproduction "too mechanical for a living organized system," and "without analogy" in nature.[40] Although Wordsworth, Barbauld, and Southey disparagingly associated Darwin's poetry with mechanism, scientific thinkers such as Sir Humphry Davy criticized Darwin's ideas about plant physiology as being instead too vitalistic and asserted that "plant movements were caused by mechanical means and that plants lacked all sensitivity."[41] Indeed, when, in 1792, the naturalist Robert Townson read a paper to London's Linnean Society titled, "Objections against the Perceptivity in Plants, So Far as Is Evinced by Their External Motion," employing mechanical philosophy to declare that Percival's vitalist theory of plant feeling should be numbered "amongst the many ingenious flights of imagination," one senses an indictment of Darwin as well.[42] In his floral clock scene, Darwin's description of anthropomorphized flowers remains

FIGURE 4. "Night-Blowing Cereus, or Cactus Grandiflorus," from Robert John Thornton's *New Illustration of the Sexual System of Carolus von Linnaeus*, 1807; 1799–1810. Notice the moon and the clock (displaying that it is just past midnight) in the background to indicate that this flower opens at night. Thornton's book includes verses from botanically related poems by Erasmus Darwin, Charlotte Smith, Anna Seward, Anna Barbauld, Robert Burns, Frances Arabella Rowden, and James Thomson, among others. (Image courtesy of the Linda Hall Library of Science, Engineering & Technology)

playful, but creates an uneasy alliance between mechanistic and organic nature that balances possibilities of their outward and inward pacemaking impulses. Charlotte Smith, publishing verses on the floral clock eighteen years after *The Loves of the Plants*, produces greater scientific distance from Linnaeus and mechanism, reevaluating debates about horological flowers' movements as demonstrating sentiency and agency, as well as Time's capacity to generate "domestic Bliss."

Smith's Horologe: Sensibility and "Conscious" Nature

If Darwin imaginatively likens the floral clock to a mechanical watch, Charlotte Smith specifically avoids mechanical analogies for horological flowers. Her posthumously published volume, *Beachy Head, Fables, and Other Poems* (1807), includes her poem "The Horologe of the Fields: Addressed to a Young Lady, on Seeing at the House of an Acquaintance a Magnificent French Timepiece." Throughout her career, Smith wrote poems displaying acute knowledge of Linnaean botany.[43] Her "Horologe" likely responds to Darwin's floral clock episode, just as her poem "Flora" rewrites *The Loves of the Plants*, retaining scientific specificity while shifting focus away from sexuality.[44] Smith's "Horologe of the Fields" reconceptualizes Darwin's floral clock by evading, for example, his irreligious exhortations and assuming an expanded perspective of botany and flowers' timekeeping. While Darwin versified Linnaeus's botanical system, Smith draws not only on the Swedish naturalist but also on several British botanists, more thoroughly exploring particular flowers of the horologe and their hours of opening and closing in Britain (not Sweden). As in her other poems about natural history, she combines precision and poetics, employing species' scientific and common names in her verse, and frequently providing the alternative denomination in her scientific prose notes.

Although Smith begins with Darwin's chief conceit, contemplating the relationship between mechanical and botanical clocks, she stresses their distinction. Rather than melding natural objects and machinery, she locates mechanism within a domestic space, the house of a wealthy young girl, that contains the manufactured object of the poem's subtitle, "a magnificent French timepiece." Here, Smith's speaker expresses the cautionary hope that

> For her who owns this splendid toy,
> Where use with elegance unites,
> Still may its index point to joy,
> And moments wing'd with new delights.
> Sweet may resound each silver bell,—
> And never quick returning chime,
> Seem in reproving notes to tell,
> Of hours misspent, and murder'd time. (ll. 1–8)

As a "splendid toy," the mechanical clock typifies a child's distraction that momentarily dazzles with "joy" and "new delights," but it also contains the possibility of failing to sustain this attention, causing time to become repetitive and dull. In this scientific poem, while the clock's "index point[ing] to" hours of potential "joy" refers to the hand on the clock, it further conjures up the "index" of a book, systematic lists cataloging knowledge apart from nature, abstracting living organisms and arguably associating (Linnaean) botanical taxonomies with this mechanical realm. Indeed, the only contemporary English translation of Linnaeus's *Philosophia Botanica*, Hugh Rose's *Elements of Botany* (1775), describes Linnaeus's list of the flowers and their times of opening as an "hour-index," essentially his blueprint for the floral clock itself.[45]

In Smith's depiction, the mechanical clock regulates its owner's behavior, "reproving" "hours misspent and murder'd time," dangers particularly acute, it seems, for those who possess luxury and wealth, and thus time to "kill." Since Smith identifies this "magnificent" timepiece as French, its threat of narrating tales of "murder" recalls that nation's violent revolution and reign of terror in reaction to social inequalities represented in aristocratic luxuries. It also conjures up the merciless, clock-like regularity of the machine administering death to social elites as, "for many, the guillotine's mechanical nature set the tone for the mechanization of death, robbing death of any honor or dignity."[46] Interestingly, in the year prior to this poem's publication, France abandoned its revolutionary calendar, which was meant to reflect natural rhythms, now indicating the failure of its "new temporal order."[47] In this poem's tense domestic space, where wealth and artifice signal dangerous connotations, the timepiece's "silver bell" and "wing'd" moments contain botanical and zoological associations that only accentuate nature's absence.

Smith's critique of abused time and luxury among the upper classes, invoking the possibility of punishment, thus generates relief when she turns away from time's manifestation in mechanical art to that which is more equally available in the natural world. She states:

Tho' Fortune, Emily, deny
To us these splendid works of art,
The woods, the lawns, the heaths supply
Lessons from Nature to the heart.

In every copse, and shelter'd dell,
Unveil'd to the observant eye,
Are faithful monitors, who tell
How pass the hours and seasons by. (ll. 9–16)

Class contentions stand out in relief as "Fortune," personifying both fate and wealth, produces autobiographical recognition of Smith's well-known announcement, in the sixth edition of her *Elegiac Sonnets*, of legal failures to award her children's rightful inheritance, a lack of "fortune" that forced her to write tirelessly to support her many dependents. In fact, an early biographer of Smith suspects that the "Emily" of this poem represents the poet's granddaughter, lending further personal relevance to the speaker's complaint of the "Fortune . . . den[ied] / To *us*."[48] Although incapable of affording "these splendid works of [mechanical] art," the speaker and Emily have *nature's* clock, by which they attain pleasures through direct botanical knowledge and make sensible use of time.[49] Nature here denotes the realm of the living "heart," providing moral guidance through "faithful monitors" more adept than mechanical "toy[s]" in educating readers of "observant eye" in "lessons" of science and sensibility. Like the "magnificent" clock, the flowers "tell" the time, but theirs is a narrative of improvement, not reproof, elucidating "how pass the hours and seasons by"—that is, what the time is, how natural processes reveal that time, and how to spend it.

Significantly, Smith's division between nature and artifice also underscores some of the contrasts between contemporary classificatory systems. Citing Linnaeus only once in her poem, Smith's subtle distancing from the Swedish botanist comes at a time when his plant theories were losing ground in scientific debates. Although many Romantic-era writers praised his natural (or authentic) and straightforward use of language, his botanical system of classification received attack as an artificial (not natural) system, basing taxonomy on sexual parts.[50] While artificial systems impose order on the world, natural systems discover some fundamental way in which it is ordered.[51] Linnaeus's sexual system "brought together plants that to the practiced eye of the expert botanist had little in common, and separated others that were very similar in most respects" (64). Linnaeus himself knew that his system was artificial, but hoped it would lead to the discovery of a natural taxonomy.[52] Indeed, in *Phytologia*, Erasmus Darwin dedicated his final section to a "plan for disposing part of the vegetable system of Linnaeus into more natural classes and orders."[53] In her natural history poetry, Smith thus employs Linnaean taxonomy while remaining cautious of its errors and mechanistic artificiality.[54]

Drawing on additional botanists, then, Smith reveals an important scientific concept operating within the floral clock:

The green robed children of the Spring
Will mark the periods as they pass,
Mingle with leaves Time's feather'd wing,
And bind with flowers his silent glass. (ll. 17–20)

Smith's image of leaves "mingl[ing]" with "Time's feather'd wing" alludes to what contemporary naturalists called "winged leaves," which were the leaves involved in the "most widely investigated" plant rhythms, coined by Linnaeus as "sleep movements" in a phrase lending itself to human analogues for plants' motions and potential sentiency as well as agency.[55] In her poem's opening note, Smith declares that "the sleep of plants has been frequently the subject of inquiry and admiration" and references Colin Milne's *Botanical Dictionary* regarding the main action of her poem, the "VIGILÆ PLANTARUM," or vigils of plants, by which "botanists comprehend the precise time of the day in which the flowers of different plants open, expand, and shut."[56] In his article on "motion," Milne describes how "during the heat of the sun in the day-time, the pinnated or winged leaves of several plants . . . rise vertically upwards," and during "the SLEEP of plants," they "after sun-set . . . hang vertically downwards, and are applied closely together, like the leaves of a book."[57] Interestingly, Smith appropriates Milne's term, the "vigils" of plants, rather than Darwin's (or Linnaeus's) easily punned "watch" of flora. Smith herself never suggests that flowers should be arranged in the shape of a clock or watch. As her title proclaims, this is the "horologe of the fields": because configuring flowers in mimicry of a mechanized timepiece would corrupt nature into artifice, these individual horologes (flowers) must be sought in their respective habitats, the knowledge of which comprises part of the speaker's educational agenda.[58]

Exploring eight "solar" flowers, "which observe a determinate time in opening and shutting," Smith constructs mini-narratives based on the flowers' names, physical traits, and natural environments. In this way, the *Nymphæa alba* or white water lily, for instance, becomes a "modest" "virgin" "cradled on the dimpling tide," while the *Ornithogalum umbellatum* or "Bethlem-star" is "pale as a pensive cloister'd nun" that responds to "vesper gales."[59] In her verse, Smith conveys flowers' times of opening and shutting mainly in relation to the position of the sun, and she specifies in her notes the precise hours of their "vigils," citing John Lightfoot's *Flora Scotica* (1777, 1789) and especially William Withering's *Botanical Arrangement of British Plants* (1787–92). Although she gestures toward this exogenous solar stimulus for plants' timekeeping, so that, for instance, the "Bethlem-star, her face unveils, / When o'er the mountain peers the Sun," she anthropomorphizes these flowers as active, rather than passive, and full of feeling and motion (ll. 42–43). In Smith's portrayal of the flowers in terms of vital, voluntary actions that privilege the presence of endogenous rhythms and faculties within these species, "the humble Arenaria," or *Arenaria marina*, thus "creeps" and "expands," as well as "sleeps" (ll. 46–48). Similarly, the *Nymphæa alba* "rests," "rises," and "sees," while the Goatsbeard "shuts" its petals and "retreat[s]," and *Silene noctiflora* "declines / The garish noontide's blazing light" and "gives all her sweetness to the night" (ll. 24–27, 38–40, 61–64). Moreover, Smith describes *Nymphæa* as "*conscious* of the earliest beam," and, in eighteenth-century materialism, *consciousness* and *voluntary* comprised keywords in determining agency (l. 25).[60] Smith's rendering of these horological flowers thereby perhaps has less in common with Darwin's mechanistic floral watch than with Percy Shelley's later, more organic sensitive plant, whose "sleep" contains "an ocean of dreams without a sound / Whose waves never mark, though they ever impress / The light sand which paves it—Consciousness."[61]

Thus, although Smith's poem ends with echoes from Darwin's floral clock episode, she provides a different moral:

> Time will steal on with ceaseless pace,
> Yet lose we not the fleeting hours,
> Who still their fairy footsteps trace,
> As light they dance among the flowers. (ll. 69–72)

These "fairy footsteps" of "fleeting hours" that "dance among the flowers" reiterate Darwin's "each light *Moment*, as it dances bye / With feathery foot." However, whereas Darwin's verses instill the primacy of science and sexuality, Smith's lines reinforce her instructive theme of science and sensibility, whereby tracing time in flowers becomes as educational, useful, and affecting as it is entertaining. In depicting the flowers' active movements and sentiency, her poem associates their timekeeping with conscious life. As "the winged moments fly" in this natural landscape, Time, with its "feather'd wing," assumes bird-like motion and vivacity. In contrast, Time appears static, stilted, and easily "los[t]" in the dead and artificial domestic timepiece of the rich girl who "kills" time because she has nothing better to do. Here, Smith references a traditional philosophical notion voiced, for example, by John Locke: "*Morality and Mechanism together* . . . are not very easy to be reconciled, or made consistent."[62] Anticipating Coleridge's complaint that mechanism "strikes Death," Smith's Lockean metaphor of a "violent Passion" in which her young acquaintance may "murder" Time implies that wealth and luxury place the girl in danger of forfeiting reason and responsibility for her actions, and of thus becoming like a machine herself.[63] Instead, describing "each flower and simple bell, / That in our path *untrodden* lie," Smith indicates that no violence is committed against time or nature in the vibrant, equalizing space of the "fields." For her, the study of nature, and of botany in particular, in which we can understand plants' pacemaking rhythms as conscious life, is never a waste of time, but vitally improves both the heart and mind.

Smith's *Beachy Head, Fables, and Other Poems* received mixed reviews from critics perplexed by her scientific acumen. One critic from the *Monthly Review* praises Smith's "tenderness and sensibility" as well as "moral reflection," but demurs that "the dry details of natural history" constitute a "pursuit [that] may seem less worthy the attention of a poet, and less calculated to excite those strong emotions in the reader which poetry should endeavor to awaken," and complains that some poems are "too technically botanical."[64] A reviewer from the *British Critic* more harshly evaluates her use of science, deploring that, "with regard to some subjects beyond her line of experience, reading, and indeed talent, [Smith] was unfortunately wayward and preposterous."[65] Although he singles out the "Horologe of the Fields" as "a very elegant and well-timed composition," he finds her notes on natural history to be "of no material value." On the one hand, this dismissal of Smith's scientific knowledge represents typical critical treatment of contemporary women writers' claims to expertise in the natural sciences outside the realm of educational texts for children (and even such pedagogical works sometimes provoked critical invective).[66] On the other hand, these reviewers' disapproval of natural history as a subject "[un]worthy the attention of a poet" because it is "dry," unemotional, and too technical repeats earlier attacks on Darwin's poetry, including those made by major male Romantic-era poets, such as Wordsworth (as discussed in this book's introduction). These critics thereby tacitly relegate Smith's verse to the "Darwinian school of poetry" through a growing critical insistence on the supposed poetic incompatibility of science and sensibility, despite her obvious departures from Darwin's work and adherence to concepts of sensibility. Such critical complaints also show, in 1807, a more general and increasing discomfort with poets' attempts to contribute seriously to scientific knowledge and its dispersal.

Smith's reduced reliance on mechanism and Linnaeus, when compared with Darwin's floral watch, exemplifies the trend within scientific thought over the following two decades.[67] Although the founder and president of the Linnean Society, Sir James E. Smith, with whom Charlotte Smith corresponded, held to Linnaean principles, viewing plants as separate from the animal realm and as displaying mere mechanical responses rather than conscious reactions, botanical debates between mechanism and vitalism continued throughout this time.[68] For instance, in 1806, the year before Smith's poem was published, Thomas Andrew Knight performed experiments showing that gravity, not volition, caused plants to grow their roots downward and shoots upward, and argued that plants did not possess free will, but instead responded to external stimuli. In 1812 James Perchard Tupper, on the other hand, published *An Essay on the Probability of Sensation in Vegetables*, which exemplified sleep in plants as indicative of sensation and instinct (169–74). At the same time, the natural method of botanical classification replaced Linnaeus's artificial system slowly in England and "in the English-speaking world, partly through pragmatism (when there is a system that works, why bother to go for something different?) and partly because of the conservatism of the Linnean Society."[69] In 1821, Samuel Frederick Gray published the first full-scale botanical work in English arguing for the natural system, writing, "Linnaeus had pronounced the discovery of the natural arrangement of plants ... to be nearly hopeless; but the French botanists ... carried it to a degree of perfection."[70] As Linnaeus's system began to lose prestige among botanists, so did it begin to fall out of fashion among general audiences. Indeed, as I further explore in chapter 6, the poet John Clare viewed Linnaean taxonomy and especially binomial nomenclature as potentially sinister and deserving of suspicion.[71] These converging factors, in combination with the growing professionalization and masculinization of the natural sciences in the early decades of the nineteenth century, influenced reformulating poetic approaches to the sciences, and to Linnaean concepts in particular, including the floral clock. In response, Felicia Hemans's versification of flowers' timekeeping dramatizes these changes by situating Linnaean botany in the contexts of antiquity and abstract meditations on human existence, confusing whether science functions as center or periphery.

Hemans's Dial: "'Twas a Lovely Thought"

Felicia Hemans published her poetic portrayal of Linnaeus's floral clock, "The Dial of Flowers," in *The Amulet; or Christian and Literary Remembrancer* (1828), in the company of works by Hannah More, Anna Barbauld, Coleridge, Clare, and L.E.L.[72] Subsequently, the poem was anthologized, sometimes with Smith's "Horologe of the Fields," in various nineteenth-century annuals (plant pun always intended), pocket books, and giftbooks collecting poetry about flowers, displaying the poet's popularity. In her collection *The Literary Women of England* (1861), Jane Williams highlights the disparateness of the floral clock poems by Smith and Hemans, remarking that "the subject is the same, the treatment as different as it could possibly receive from two feminine minds."[73]

While eighteenth-century naturalists often encouraged the participation of nonspecialists, this practice slowly changed with science's increasing professionalization in the early decades of the nineteenth century, and shifts in literary styles and subjects in many ways reflected a growing difficulty in claiming scientific authority in serious, published poetry. Although Hemans demonstrates botanical knowledge, she most obviously aligns her "dial of flowers" with myth and religion. As Gary Kelly states, "Hemans did not relinquish ambitious and potentially 'unfeminine' subjects [such as science], but rather framed them in ways that would seem acceptably feminine."[74] Known as "Mrs. Hemans," she famously cultivated a feminine and domestic persona based on sentiments of the heart.[75] Unlike the more caustic critical reactions to Smith's scientific poetry, Hemans's reviewers describe her works as "always welcome" and her name as one "the eye rests upon with delight," deeming her "Dial" "an exquisite little poem."[76] Titling her poem "The *Dial* of Flowers," rather than punning on the "watch" of flora or flower "clock," Hemans avoids these modern mechanical, artificial devices in favor of the sundial's more primitive technology to analogize vegetable horologes and their use of light.

In contrast with Smith's many scientific notes, Hemans's poem contains only one note, explaining of the *Horologium Florae*, "This dial was, I believe, formed by Linnaeus, and marked the hours by the opening and closing, at regular intervals, of the flowers arranged in it." Hemans's hesitant "I believe" when attributing the floral clock's origins to Linnaeus presents him now as somewhat obscure, his ideas less known, and produces a poetic persona claiming only casual knowledge of his botanical influence. Her poetic distancing of both herself and Linnaeus from serious science also permeates her poem's first line, creating the challenge of understanding Hemans's subtle participation in botanical debates earlier associated with the floral clock. Her first three stanzas illustrate this formulation of scientific concerns in unassuming, imaginative terms:

'Twas a lovely thought to mark the hours,
As they floated in light away,
By the opening and the folding of flowers,
That laugh to the summer's day.

Thus had each moment its own rich hue,
And its graceful cup and bell,
In whose color'd vase might sleep the dew,
Like a pearl in an ocean-shell.

To such sweet signs might the time have flow'd
In a golden current on,
Ere from the garden, man's first abode,
The glorious guests were gone. (ll. 1–12)

Opening with the phrase "'Twas a lovely thought," Hemans undermines the floral clock's scientific legitimacy with the designation of "lovely" and relegates this Linnaean concept to the past. Her dismissive move, so seemingly innocuous, enables her location of Linnaeus's botanical systems and "thought[s]" within distant myth and spiritual, idyllic simplicity. Employing the idea of the floral clock but seemingly discarding its science, Hemans casts Linnaean mechanism as outdated and irrelevant, and Linnaeus himself more as a dreamer than a theorist of the natural world. At the same time, while disclaiming scientific knowledge serves as a modesty trope for the female poet, Hemans's alignment of Linnaeus with imagination also paradoxically creates a space for her engagement with his botanical ideas without appearing to encroach on the (masculine) professional territory of science. Unlike Smith's, Hemans's nontechnical botany consequently appeals to contemporary readers and critics as unthreatening within the realms of the feminine and poetic. Thus, her efforts to downplay science also draw attention to its implied presence within the poem.

Throughout her initial three stanzas, Hemans employs words and phrases that evoke specific botanical ideas while leaving this expertise unexplored. For instance, her reference to flowers' timekeeping systems functioning "in light" conjures up botanical processes including photosynthesis, which was first fully described by Erasmus Darwin and constitutes "the most usual and important synchronizer of circadian rhythms" (l. 2).[77] Moreover, Hemans's floral synecdoche of "cup and bell" highlight Linnaean terms with double meanings, for, as the most recent English translator of *Philosophia Botanica* explains, "some words that [Linnaeus] regularly uses in a technical sense can also appear with a more general application; thus *pistillum* can be the pistil of a flower or the clapper of a *bell; calyx* the botanical calyx or a common *cup*" (ll. 6, 5).[78] In translation, Linnaeus's Latin terms blur distinctions between the scientific and the mundane, a difficulty similar to that of Hemans's poem. Evoking contentions about plants' sentiency and motion, her flowers feelingly "laugh," and the word "sleep" easily references Linnaeus's "sleep of plants" that Smith's notes to "Horologe" discussed in detail (l. 7). This scientific connotation becomes even more probable since Hemans depicts it in conjunction with "dew," a crucial component in triggering the "sleep" or motion of plants' leaves, which, as Milne delineates, "can be produced by an artificial as well as natural dew."[79] Additionally, when Hemans states that time "flow'd / In a golden current" to these flowers, associating the floral clock with the prelapsarian Garden of Eden, this "golden current" at once signifies this golden age and scientific interpretation through the "golden current" of electricity (l. 10). Naturalists, including Erasmus Darwin, wrote of electricity as a force promoting growth in plants and controlling their natural rhythms. In *Phytologia*, Darwin records an ongoing experiment in which one naturalist "for the purpose of keeping a few flower-pots perpetually subject to more abundant electricity ... affixed a small apparatus to the pendulum of a clock."[80] Hemans thus participates in botanical debates and discourse even as she dilutes these scientific allusions with myth and metaphor.

Conjecturing that the floral clock kept time in various pastoral paradises, including the heathen afterlife of Greek myth, Hemans then changes tone in her final two stanzas (l. 17). She moves away from the ideal, inciting moral application of the flower dial to mortal experience:

> Yet is not life, in its real flight,
> Mark'd thus—even thus—on earth,
> By the closing of one hope's delight,
> And another's gentle birth?

> Oh! let us live, so that flower by flower,
> Shutting in turn may leave
> A lingerer still for the sunset hour,
> A charm for the shaded eve. (ll. 21–28)

Hemans here takes the floral clock out of myth and transplants it in the "real" "life" of individuals, conveying simultaneously general and unique rhythms of humanity. Analogizing the dial with cycles of human hope, she depicts each flower's period of opening as successive dreams or goals within an individual life, each "shutting in turn." A single day-length of the flower dial becomes a full human existence, punctuated by the wish that one final hope (or flower) will remain blooming into the hour of death (or sunset). She thus transforms the floral clock and its scientific associations into a message of consolation for human mortality.

Hemans's nontechnical rendering of the floral clock thereby aligns more closely with poetic representations of scientific ideas in what had by this time become the "high" literature of contemporary writers such as Wordsworth and anticipates the value of spiritual, sentimental verse in the Victorian age.[81] Indeed, Hemans viewed Wordsworth as her poetic mentor, often adapting his subjects and style of verse.[82] As opposed to the anthropomorphized scientific flowers of Darwin and Smith, Hemans does not name or personify any individual flowers, and her conclusion differently analogizes the floral dial with human feeling and experience, resembling Wordsworth's approach to flowers' potential for emotional intensity through time. When, in his earlier "Ode: Intimations of Immortality" (1807), Wordsworth famously gestures toward a common flower as a rejuvenating force that reappears each spring and produces temporal dissonance for the speaker who associates it with his youth, he laments that, while the landscape seems the same, he himself has changed, so that "the meanest flower that blows can give / Thoughts that do often lie too deep for tears." For him, the flower's conjunction of time with movement and feeling culminate in his own response—the speaker feels moved—rather than in consideration of the flower's possible sentiency or movements. In such poems by Hemans and Wordsworth, scientific conceptions of the floral clock thus provide a framework for flowers' signification of time, "lovely thought[s]" that inform their more central explorations of human experience and emotion, marking important shifts in the relationship of Romantic-era science and poetry.

Moving from conceptions of temporality, my next chapter discusses sensibility in the context of botany and empire. While James Grainger here provided insights into how a floral clock might be constructed within the West Indies through his georgic poem focused on those colonies, the poet and travel writer Maria Riddell instead employs botany both as a means to communicate the knowledge and actions of less-often acknowledged Caribbean inhabitants, and to subtly (and, in some ways, conflictedly) elicit sympathy for those communities from British readers. Additionally, although I have shown within this study that Erasmus Darwin's botanical poetry received critique for being devoid of feeling, Riddell's 1802 edited collection of poems includes works by Darwin that display his ability to write verses more obviously reflecting emotional concerns within the culture of sensibility, thus complicating critical perspectives of his oeuvre.

PART II

Sensibility and Empire

GENDER, RACE, AND NATION

3

Transformations of Gender, Race, and Poetic Sensibility

Maria Riddell's Transatlantic Botany and Biopolitics

Debates about the fixity or dynamism of race, sex, and nationality fascinated naturalists in the second half of the eighteenth century. The prefix *trans-*, meaning "from one place, person, thing, or state to another," was applied in various ways, perhaps most importantly for this chapter in the context of the "transatlantic."[1] In this period, ideas about biological transformations became particularly associated with European nations' transatlantic colonies. Scientific thinkers such as the French naturalist Georges-Louis Leclerc, comte de Buffon linked the Americas and the West Indies to concepts of hybridity as well as degeneration (meaning a loss of size, strength, and reproductive abilities) from "original" biological forms found in the Old World and, especially, Europe.[2] Such potential for alteration raised questions about the effects of transatlantic movement on living organisms, including humans. In various ways, British writers such as Janet Schaw, William Smellie, Erasmus Darwin, John Clare, and Maria Riddell show how theories from natural history extended to notions of transformation regarding nation, sex, or race in Britain and the Caribbean islands.

Riddell's *Voyages to the Madeira, and Leeward Caribbean Isles: With Sketches of the Natural History of these Islands* (1792) records her observations of West Indian natural history, particularly in Antigua and St. Kitts, and the knowledge practices of white women colonists and enslaved African laborers, two groups that often signaled concerns about hybridity and transformation in the British imagination. Riddell thereby participates in larger contentions about the stability of these identities in the context of empire.[3] Scientific ideologies of the late eighteenth century generally upheld Enlightenment ideas of determinism and sociobiological fixity in sex, race, class, and nation, claiming evidence through naturalists' observations (and impositions) of taxonomic order, behavior, and physiology in plant and animal species. Many naturalists interpreted change and ambiguity in natural forms as signs of anomaly or monstrosity. While masculine European traditions of knowledge singled out white women and Black people of all genders to typify particular kinds of degeneration, Riddell suggests that these denigrated groups possess valuable insights into the natural world, and she incorporates subtle cues of sensibility to direct British readers' sympathies toward these West Indian colonial counterparts. Through her travel narrative, personal letters, and poetry, she engages and challenges naturalists' theories about hybridity and transformation in botanical and zoological forms as well as the resulting social implications for inhabitants of the West Indies, particularly questioning the direction in which perceived improvement occurs in these colonies. Moreover, *Voyages* offers a rare account of Britain's Caribbean colonies at a time of monumental political transition relating to the American, French, and Haitian revolutions. Within this framework, Riddell registers limits as well as possibilities for reimagining gender, sex, race, and nation in her contemporary society and helps to redefine transatlantic capacities for change.

In the chapter's final section, I analyze selections from Riddell's poetry about the West Indies, among other subjects, and especially consider her edited collection, *Metrical Miscellany* (1802), which includes poems not only by Riddell but also by authors such as Anna Barbauld; Georgiana, Duchess of Devonshire; and Erasmus Darwin. Darwin's poems become especially important because, although his contemporaries often accused him and his botanical verses of lacking feeling, his poems in Riddell's volume display his ability to write within more emotional registers of sensibility, thereby complicating those critical perceptions of his works. In these ways, Riddell's transatlantic oeuvre also crucially expands understandings of Romantic-era scientific, literary, and (bio)political networks and some of the individual participants in Britain and the Caribbean colonies.

"Universal Anarchy": Sociobiological Alterations through Plants' "Sexual System"

Embodying a British national hybridity, a theme of her travel writing's depictions of colonial identity, Riddell was born in England but had strong familial ties to the West Indies.[4] Her father, William Woodley, twice served as governor of the Leeward Islands; her mother, Frances Payne, was "the only surviving daughter and heiress of Abraham Payne of St. Kitts," and Riddell was married on St. Kitts in 1790 before moving to Scotland.[5] In *Voyages*, Riddell displays a fascination with conceptual and biological hybridities in species found in the West Indies, such as the mule, "sea bat," and "animal flower" (53, 55, 68). A poet as well as a travel writer, Riddell was acquainted or became friends with notable contemporaries, including Anna Barbauld; Erasmus Darwin; Henry Fuseli; Walter Scott; Matthew "Monk" Lewis; Georgiana, Duchess of Devonshire; Henry Erskine; and Robert Burns. Indeed, it is her role as friend and memoirist of the Scottish bard for which Riddell currently is best known. Crucially, Burns introduced her to his publisher, William Smellie, who was also a naturalist, functioning as the keeper and superintendent of Edinburgh's Museum of Natural History from 1781, authoring *The Philosophy of Natural History* (1790), and translating into English Buffon's works of natural history. In addition to publishing her *Voyages*, Smellie became Riddell's friend and correspondent, often discussing with her topics related to their mutual interests in the natural sciences. In a later letter, imagining her traveling and "botanizing on the Alps," he wrote, "When you go to the continent, besides your common researches concerning plants and animals, I hope you will not neglect proper observations on that strangest of all animals, Man."[6] In response, she cleverly described "that species of animal called a Traveller," and, as Smellie knew, and as is apparent in *Voyages*, she often employed her examinations of botanical and zoological species as a means also of focusing on "human life" (Kerr, 2:389).[7]

Like many other outspoken women of wit and education in her time, Riddell came under attack, especially for her interests in science and politics. Mary Wollstonecraft, a travel writer herself, represents perhaps the era's most famous example of a woman's intellectual assertions being interpreted as sexual transgression.[8] In this vein, one of Riddell's most caustic critics, Charles Kirkpatrick Sharpe, describes her as "a worthless profligate woman" who was "shunned by her own sex."[9] A letter from Burns jocularly indicates her social difficulties, when, advising her to use her influence for a friend's advancement, he remarks, "I was going to mention some of your female acquaintance, who might give you a lift, but on recollection your interests with the WOMEN is I believe a sorry business," playfully blaming this circumstance on her "despotic use of ... sway over the Men."[10] Riddell wrote in a private letter, "There is but one animal I think more inconsistent, more fickle, less to be trusted, and with a lesser remaining impress of the Creator's stamp about it, than Man—I mean Woman" (Kerr, 2:391). She expressed the desire "to free" her daughter "from the little weaknesses and delicacies that render women 'interesting'—and miserable, 9 times out of 10" in a Wollstonecraftian challenge to conventionally confining gender roles (2:391).[11] When Horace Walpole famously called Wollstonecraft a "hyena in petticoats," his insult evoked the common belief that hyenas could change sex, suggesting that this writer and defender of *A Vindication of the Rights of Woman* had to become "male" in the process.[12] Although Riddell's thoughts on Wollstonecraft's treatise are not extant, she admired William Godwin's *Political Justice* (1793), advocated women's education and, as we shall see, similarly was classified as a natural object during an incident in which her mentor Smellie accused her of attempting to change sex within the context of her scientific knowledge.

According to most eighteenth-century naturalists, the capacity for biological alterations or transformations from original forms largely depended on hybridity, which could occur through the sexual union of two different species.[13] As a concept, hybridity sparked naturalists' debates over analogical possibilities between social and natural orders. This controversy had obvious import for British colonists in the West Indies, where frequent hybridity as well as physiological changes in transported botanical and zoological species were widely acknowledged. As I have discussed, Linnaeus's "sexual system" of botany, which classified plants according to their reproductive parts and placed them within "marital" relations, enjoyed immense popularity and, especially as versified by Erasmus Darwin, emphasized analogies between plants and humans. In Darwin's depiction, hybridity guides Linnaeus's account of the origin of species; Darwin explains that "one plant of each Natural Order was created in the beginning; and that the intermarriages of these produced one plant of every Genus, or Family; and that the intermarriages of these Generic, or Family plants, produced all the Species: and lastly, that the intermarriages of the individuals of the Species produced the Varieties."[14] Indeed, for Darwin, the formation and progress of both zoological and botanical species often result from hybridity.¹⁵ However, Buffon and his followers—Smellie in particular—advocated comparisons between humans and zoological species, deriding the sexual system of plants as a misapplied analogy. Buffon, citing the production of mules, asserts that hybrids occur most frequently in warm climates, noting a prevalence in the West Indies (8:15–17). While Buffon assigned vegetation (particularly as a food source) a role in effecting alterations in animals and humans, he rejected Linnaeus's attempts to draw comparisons between plants and humans. Buffon instead elevated his own work on zoology as the surest means of gaining insight into both humankind and the potential for hybridity.

As many critics have argued, Buffon first used the concept of reproduction in its modern sense, which "was also deeply connected to the development of ideas of race and sexual difference," as well as "to the emergence of biopolitics as a modern phenomenon."¹⁶ Attempting to control biopolitical implications, in *The Philosophy of Natural History* Smellie ridicules Linnaean analogies in the "sexual system" between plants and animals or humans, anxiously imagining the hybridizing "consequences" that such sexual reproduction would imply for the vegetable kingdom. Smellie argues that, in plants, pollen

> by flying *promiscuously* abroad . . . might impregnate different species which happened then to be in a fit condition for the reception of male influence . . . Nature intends that plants should multiply and perpetuate their kinds; but the sexual hypothesis makes her take the most effectual measures to prevent that intention, and to introduce *universal anarchy* among the vegetable tribes. Were [the Linnaean sexual system] true, the whole vegetable kingdom, in a few years, would be utterly *confounded*: Instead of a regular succession of marked species, the earth would be covered with *monstrous productions*, which no botanist could either recognize or unravel.¹⁷

Smellie dismisses Linnaeus's sexual system of botany and assures that, contrary to such "anarchy," "all laws of Nature are fixed, steady, and uniform" (251). Although not discussing Smellie, Jill H. Casid writes that "eighteenth-century use of the adjective *promiscuous* and the adverb *promiscuously* already made implicit and explicit analogies between human sexual behavior and the environment."¹⁸ While Smellie challenges Linnaeus, especially by portraying pollen as "flying promiscuously abroad," the sexual language of plants also "provided a way to speak of what was considered taboo interracial sex" even as it "endeavored to displace and confine the romance of the Other to plant life"—a romantic prospect that Smellie found discomfiting, to say the least.[19]

In Smellie's formulation, uncontrolled plant sexuality disrupts order and continuity in a way that, if allowed to be analogical with humans, threatens sex, class, and racial boundaries. For him, in a sexual system of vegetation, the production of hybrid monstrosity hinges on the (female) plant's indiscriminate receptivity to (male) pollen, regardless of "kind." It is on female sexual control that "kingdoms," both national and natural, therefore depend for their preservation.[20] Following the same logic, the racial phobias of Edward Long, a contemporary historian of the West Indies, center on the behavior of women in his warnings about the concurrent scene in London, where, he claims, "the lower class of women in England, are remarkably fond of the blacks [and] . . . in the course of a few generations more, the English blood will become so contaminated with this mixture . . . this alloy may spread so extensively, as even to reach the middle, and then the higher orders of the people," altering "the whole nation."[21] An advocate of slavery and notions of polygenesis, Long drew the botanical and zoological analogy, claiming that unions between Black and white "species" resulted in a monstrosity of "contaminated" hybrids or "mixed progeny."[22] For Smellie, perhaps more than race and nation, class and especially sex and gender are clear subjects of his concern.[23] In *Philosophy*, he apologizes for a discussion of "hermaphrodite plants," parenthetically explaining that he "must speak in the language of the system"; he attacks the idea of spontaneous sex changes asserted by Linnaean "sexualists," through which "trees, which had continued many years under the character of females . . . had suddenly dropped their female forms, and assumed the more robust features peculiar to the male part of the creation!" (252). Smellie's emphatic resistance to ideas of sexual transformation and freedom in both social and botanical spheres helps to show that "hybridization is not so easily disentangled from its colonial legacy in the founding discourses of botany and race," with implications for sexuality as well as for imperial power.[24]

While Riddell "set so high a degree of value and esteem upon" Smellie's *Philosophy*, she chastised his efforts "to controvert [Linnaeus's botanical] systems" and to confine expansions or transformations of gender and sex (Kerr, 2:380). In her *Voyages*, while nationalistically employing the zoological classifications of the British naturalist Thomas Pennant, Riddell explains that she "always made use of [Linnaeus's] generic and specific names, because they are now the most universally received by naturalists," and provides the Linnaean names for botanical species whenever possible as well (vii–iii). Indeed, she declared to Smellie, "it is to me a sacrilege to differ [with Linnaeus]" (2:380). Endorsing Linnaean analogies between plants and humans, she engages with the dynamic possibilities for gender and sex in the natural world. As Darwin displays in the preface to *The Loves of the Plants*, Linnaeus's sexual system reveals potential for humanistic metaphors ranging from the socially conventional to forms of sexual variation and fluidity within the vegetable kingdom; for example, some flowers could have a female (pistil) or "wife" with one, two, three, four, or more males (stamens) or "husbands," while others represent "feminine males," exhibit the possibility for "polygamy," and so on. As various scholars have shown, Linnaeus's sexual system of botanical classifications offered women the opportunity to explore gender and sexuality while demonstrating scientific acumen.[25]

Drawing on Linnaean plant analogies and their implications for gender and sexuality in a letter to Smellie, Riddell cleverly compares herself to a "creeping plant" in Edinburgh's botanical garden. She quips that the intensity of her studies makes her feel "rooted to one spot of the earth, and with the mere privilege of ambulating backwards and forwards on my own grounds, which is no more than" the radius of movement enjoyed by that "vegetable" (Kerr, 2:370). Policing gender lines, Smellie indulges Riddell's Linnaean analogy insofar as to inform her "that you are a *vegetable*, as you say, I allow. But that you belong to the *cryptogamia* class [in which sex is difficult, if not impossible, to determine], I deny" (2:378). Again displaying his discomfort with blurred sexual identities, Smellie strictly categorizes Riddell as female within the plant system and attempts to undermine the seriousness of her pursuit of masculine scientific knowledge, assuming a tone of stern (albeit somewhat playful) masculine authority: "I allow"; "I deny." In reply to his botanical classification, rooting her firmly in the feminine, Riddell undauntedly exploits Smellie's fears of ambiguity and hybrid monstrosity, retorting, "You ... omit making known to me under what order I am to look for *you* in the botanical dictionary; I am inclined to insert you, in the appendix to mine, as a *non-descript*" (2:381).[26] Riddell here refers to the section of "non-descript" botanical species at the end of her *Voyages*, where she notes that "the Linnaean Names of the following Plants are Unknown" and closes her work with a passage from Thomson's *Seasons*: "Thus spring the living herbs, profusely wild, / O'er all the deep green earth, beyond the pow'r / Of botanists to number up their tribes" (105).²⁷ Like many contemporary women writers, Riddell employs taxonomic ideologies even as she questions their adequacy to fully encompass multiplicities within the natural world and their sociological connotations.

In *Voyages*, Riddell selectively draws on the major naturalists of her day, retaining a complex perspective of biological alterations in the West Indies. Smellie's rejection of the sexual system of plants indicates a deeper fear of the chaotic potential of hybridity through sexual reproduction. He does not deny that "hybrids" exist in nature, but follows Buffon (whose works he translated in 1780–85), declaring that, in the vegetable world, these "variations" come about by means other than sexual procreation. For Smellie and Buffon, "culture," soil, and climate are sources of alterations in both plants and animals. Yet, significantly, regardless of the cause for botanical alterations, in the West Indies such changes in vegetation appear to reverse the generally accepted direction of biogeographical "improvement." Whereas Buffon primarily associated the West Indies with zoological degeneration, the Scots travel writer Janet Schaw, who visited the Leeward Islands less than two decades before Riddell, notes that many West Indian "plants were of the same tribes at least with what we have [in Britain], but so greatly improved, that they were hardly to be known. How different is that from the plants of this country [the West Indies], when they come to our Northern Climate."²⁸ And Riddell similarly, favorably, compares Caribbean vegetation with that of Britain and some of its frequent imports, asserting, for instance, that the manchineel tree's wood is superior "to the best mahogany" and "the *Mammoea Americana*, or *Mammee sappota*, equals, if not surpasses the English oak in beauty" (91, 96). If plants could be said to improve in the West Indies, what are the implications of this improvement within Linnaeus's botanical system of human analogues? Riddell's comparisons seem to provide the possibility for improvement (rather than degeneration) among Britons who relocate to West Indian colonies.

Schaw expands on this vegetable analogy, indicating that plants (and people) improve when transplanted from Britain to the Caribbean and degenerate when moved in the opposite direction, toward Europe. She writes of a West Indian colonist who humorously identified himself as a Caribbean plant and "told us that in compliance with the wishes of his children, he had resided in England for several years, 'but tho' they kept me in a greenhouse,' said he, 'and took every method to defend me from the cold, I was so absolute an exotick, that all could not do, and I found myself daily giving way, amidst all their tenderness and care; and had I stayed much longer,' continued he, smiling, 'I had actually by this time become an old man'" (105). In keeping with Riddell's and Schaw's botanical observations, and contrary to Buffon's zoological assertions, Britain and Europe become associated in this account with degeneration, while the Caribbean climate may represent health, strength, beauty, and improvement.

"The Rights of Half Mankind": Transformation, Women, and Slavery

In her descriptions of botanical species in *Voyages*, Riddell elaborates on plants' significance and uses for white women colonists, enslaved Africans, and the islands' Indigenous communities, providing insight into the habits of these less-often-recorded populations as well as subtle commentaries on race and gender in the Caribbean colonies. Resembling Mary Wortley Montagu's attention to women and to aspects of foreign societies to which women have privileged access in *Turkish Embassy Letters* (1763), Riddell's text reveals how women in these British colonies applied their knowledge of natural objects. For example, she explains the medicinal and domestic uses of plants such as the physic nut bush and cashew tree that contain juice "often employed to stain the initials of a name in linen" (82–83). Describing a plant cosmetically applied by women, the "*cactus*, or *prickle-pear* species," she indicates that the "West India ladies employ [its crimson juice] not only as a dye for their ribbons and gauzes, but also as one for their cheeks" (86). Caribbean plants contribute to household chores as well as to personal beauty and enhancements, both in skin applications and in accessories, as she makes clear in her description of wild liquorice, which "bears berries so beautiful, that West India ladies have them perforated, and wear them strung like beads in necklaces and bracelets," while the yellow nickar "likewise bears a beautiful berry like polished marble, worn in earrings, and as tassels to watch-chains" (90). These women thereby exemplify both West Indian botany's practical uses and its aesthetic allure.

In contemporary European imaginations, Riddell's account of white women converting berries into fashionable accessories in the West Indies also may have invoked stereotypes associated with the islands' Indigenous peoples or enslaved African population. However, this very potential for producing identification and sympathy through shared cultural practices arguably displays one of the ways in which she incorporates subtle gestures of sensibility into her text; for her, these different communities' closeness to and knowledge of the natural world assumed a positive capacity for understanding its uses, powers, and remedies.[29] At the same time, it is important to acknowledge, as Misty Krueger states, "that the freedoms afforded to some women travelers in this era, especially those of white European descent, were the result of imperialism, colonization, and Black women's trauma."[30] While transatlantic stories such as Riddell's exhibit "the transformative power of travel in terms of women's observations and examples of personal autonomy, we would be remiss not to recognize that this independence is contingent upon the involuntary migration of other women."[31] Moreover, as Shelby Johnson insightfully asserts, while white women travelers could usefully employ aspects of sensibility, nevertheless "sympathy, philanthropy, and antislavery—the hallmarks of both benevolent colonialism and domesticity—remained entangled in forms of patriarchy."[32]

In this vein, relating another instance of a plant's cosmetic application, Riddell displays how women's interactions with the natural environment could conjure up the importance of race and transformation on these islands. She explains that the cashew nut "is enveloped with a thin shell, that contains an oily inflammable fluid, which is caustic. The ladies in the West India Islands make use of it to extract freckles from their faces. They sometimes spread it all over their hands, necks, and face; and, in a few days, the skin peels off in great flakes, after which the complexion appears for some time exquisitely fair, but is more liable to sun-burn than ever; besides the pain of this operation is excruciating" (82). In Riddell's description, white West Indian women's desire to regain or maintain a fair complexion recalls to mind the effects of the sun and climate, as well as anxieties over sustaining distinctions of skin color in this slave society, stratified by race.[33] However, noting similar practices in earlier West Indian texts, Deirdre Coleman explains that "the problem with complexion is that it is an unstable boundary marker" so that such whitening is "liable to be interpreted as proof not of [white women colonists'] difference from black slaves, and from black women in particular, but of their uncanny resemblance to them."[34] As Katy L. Chiles has shown, race "was thought to be an exterior bodily trait, incrementally produced by environmental factors (such as climate, food, and mode of living) and continuously subject to change."[35] Buffon, expressing his Eurocentrism, wrote that "if, by any great revolution, man were forced to abandon those climates which he has invaded, and to return to his native country, he would, in the progress of time, resume his original features, his primitive stature, and his natural color. But the mixture of races would produce

FIGURE 5. *Cashew Tree and Butterfly Metamorphosis*, Maria Sibylla Merian, ca. 1705, from *Metamorphosis insectorum surinamensium* (Amsterdam: G. Valck, 1705), plate 16. (Image courtesy of the Huntington Library, San Marino, California)

this effect much sooner," and "150, or 200 years, are sufficient to bleach the skin of a Negro" (7:394).[36] In his own natural history text, Oliver Goldsmith likewise asserted, "All those changes which the African, the Asiatic, or the American undergo, are but accidental deformities, which a kinder climate, better nourishment, or more civilized manners, would, in a course of centuries, very probably remove."[37]

In Riddell's work, by employing the juice of an exotic plant, these colonial West Indian women strive to undo the climate's skin-altering effects in a painful endeavor that ironically makes them more susceptible to burning and darkening; nevertheless, critiques of such West Indian practices and their implications for race contrast with some contemporary portrayals of this feminine procedure of exfoliation as a potentially desirable transformation. In her illustrations of Surinam, Dutch travel writer Maria Sibylla Merian (1647–1717) depicts the cashew tree and nut in conjunction with a butterfly metamorphosis. In Merian's accompanying description of the tree, she mentions the method of applying the plant's juice to peel away skin, and, as Londa Schiebinger demonstrates, this writer was very aware of how Indigenous, enslaved African, and European colonist populations employed Caribbean vegetation to effect changes in and on the body.[38] Although Merian studied insect metamorphosis more generally, she here combines these botanical and entomological images, arguably signifying an aesthetically positive transformation available to white women through cashew-nut oil. Riddell characterizes the result of this exfoliating process as "exquisitely fair" even as she cautions that "this operation is excruciating," leaving it unclear as to whether her observation suggests a tone of judgment or the possibility of first-hand experience.[39]

Writing eighteen years prior to Riddell, Schaw employs botanical analogies to scoff at West Indian white women's efforts to avoid skin darkening. She laments that these women "want only color to be termed beautiful, but the sun who bestows such rich taints on every other flower, gives none to his lovely daughters; the tincture of whose skin is as pure as the lily, and as pale."[40] Schaw explains that this is due to "the way in which they live, entirely excluded from proper air and exercise," evading the sun by being "covered with masks and bonnets, that absolutely make them look as if they were stewed."[41] In contrast, she expresses pride in her own "brown beauty," asserting, "I have always set my face to the weather; wherever I have been."[42]

Yet it was not just in the West Indian colonies that white women applied botanical exfoliants to their skin in an effort to maintain a fair complexion and prevent darkening from sun exposure; such procedures prevalently existed in Britain as well. In fact, in Darwin's medical text *Zoonomia* (1794–96), he specifically pathologizes British white women's beauty practices through cosmetics that risk "destroy[ing]" their health due to inclusion of ingredients such as "white lead," and thus categorizes excessive pursuit of such beauty treatments as a kind of insanity.[43] Nevertheless, he provides guidance regarding how to achieve skin exfoliation at home and, rather than the Caribbean's cashew-nut oil, he explains that "the tan of the skin occasioned by the sun may be removed by lemon juice evaporated by the fire to half its original quantity," clarifying that "freckles lie too deep for this operation" (2:375). Despite elucidating this practice,

Darwin advises British women (in a manner reminiscent of Clarissa's speech in Pope's *Rape of the Lock*) that the only real "cure" for the madness of pursuing such beauty enhancements "must be sought from moral writers, and the cultivation of the graces of the mind, which are frequently a more valuable possession than celebrated beauty" (2:376).

Moreover, it was not only middling- and upper-class British women who sought these botanical methods of skin exfoliation, for it occurred among the rural laboring classes as well. In John Clare's poem *The Shepherd's Calendar* (1827), a passage from "May" discusses the common fumitory plant in this context:

> And fumitory too—a name
> That Superstition holds to fame—
> Whose red and purple mottled flowers
> Are cropp'd by maids in weeding hours,
> To boil in water, milk, and whey,
> For washes on a holiday,
> To make their beauty fair and sleek,
> And scare the tan from Summer's cheek.[44]

Thus, English women within the rural peasantry employed their botanical knowledge for these cosmetic purposes as well. Herbals had noted uses of fumitory for treatment of skin conditions for centuries. The Linnaean botanist Thomas Green wryly wrote in his *Universal Herbal* (1820) that "there is a disorder of the skin ... thought to place the empire of beauty in great jeopardy; the complaint is frequently brought on by neglecting to use a parasol," but "the infusion of the leaves of [fumitory] is said to be an excellent specific for removing these freckles, and clearing the skin," and he echoes the Darwinian warning that this "ought, we think, to be chiefly employed by those who have previously removed those moral blemishes, which deform the mind."[45]

Riddell's description of the cashew tree aligns with many other instances in her travel narrative in which the Caribbean environment provides both a danger or transformative potential (in this case, skin damage or darkening due to the climate/sun) and a possible remedy or reversal (skin exfoliation). In 1756, writing about West Indian settlers' cosmetic use of cashew-nut oil, Jamaican doctor Patrick Browne compared his women patients' casting off of skin to the shedding of skin by "snakes and adders."[46] As European travelers sometimes described Caribbean islands in terms of an Edenic paradise, reference to snakes easily evokes the biblical serpent's temptation of Eve with knowledge, power, and experience, a story to which Riddell also arguably alludes in her depiction of the manchineel tree: "Its verdure, its foliage, and the beautiful apple it bears, would render it the most delightful tree within the Tropics, were it not perhaps one of the most destructive" (91). She notes that the tree's bark is filled with a poison fluid, the very touch of "the leaves, will blister your skin," and "the manchineel apple is certain death, if eaten" (91).[47] However, if this poison apple is reminiscent of the tempting fruit in Eden, Eve's fall into knowledge and experience assumes redeeming value as signified in Riddell's scientific acumen, which can reverse this apple's detrimental effects. She reveals that another botanical species, the Indian arrowroot plant, in addition to being "a remedy in dysenteries and many other Tropical complaints," is "an effectual antidote against the venom" of the manchineel tree (95). Communicated through Riddell, this women's knowledge, rather than delivering death, destruction, and degeneration, advances a scientific means of providing medicinal healing and potential "improvement."

For some writers, the specter of temptation and sexual experience loomed large in these British colonies, bringing the possibility of racial hybridity and national transformation.[48] In her account of the Leeward Islands, Schaw wrote of West Indian women colonists that she "never admired [her] own sex more than in these amiable creoles," describing them as "modest, genteel, reserved, and temperate" (113).[49] On the other hand, while she presents creole men as gallant, handsome, "frank, open, generous," and brave, she also portrays them as "victims" of the climate and of the "young black wenches" who "lay themselves out for white lovers" (111–13). She perversely blames African women for the "widespread miscegenation that so threatened the integrity and coherence of Britain's colonies."[50] Always fearful of racial hybridity or intermixing, Long urged British West Indian men to resist such temptations and to "perform the duty incumbent on every good citizen, by raising in honorable wedlock a race of unadulterated beings."[51] In her travel narrative, Riddell approaches race and the institution of slavery somewhat differently.

Discussing "the fall of the planter class in the British Caribbean," Christer Petley recently contrasted the travel writing of Schaw (1770s) to that of Maria Nugent (1802) to show that while earlier British travelers to the West Indies "were frequently forgiving of white creole difference" and flaws, this attitude changed during the 1780s and 1790s owing to the results of the American and Haitian revolutions, the abolitionist movement, and political and economic policies that caused examples of "transgressive" colonial behavior to be viewed as "something more sinister."[52] Riddell's travel writing becomes all the more valuable as an account temporally situated between the texts of these two women, helping to reveal this transition in British perceptions of the West Indian colonies' inhabitants and changing slave society. In the same letter in which Riddell wittily wrote to Smellie that, through her studies, she assumed a "vegetable" existence, she asserts that in the West Indies yellow fever "has proved nearly as fatal, for the time at least, as the democratic spirit of equality which is spreading like wild-fire amongst the colonies, and occasioning the most terrible depredations" (Kerr, 2:274–75). Referring to a series of rebellions organized by African freedom fighters, she continues, "Our ancestors, when they instituted the accursed traffic of the slave trade, brought over a nation, who, though long patient and submissive to servitude, seem now to have nearly touched, by the decree of Providence, the term of their bondage, and have already begun to retaliate the injuries imposed upon them by their persecuting masters" (2:274–75). Although she expresses sympathy and support for enslaved peoples' plight, she regrets that violence must accompany the pursuit of freedom, adding, "But we must deplore, however zealous in the cause of liberty and justice, that the laws of humanity should thus be violated, before the rights of half mankind can be firmly established, which I fear they will not be yet without the effusion of more blood" (2:274–75).

While her family owned slavery-based West Indian sugar plantations, Riddell, who harbored strongly abolitionist sentiments, largely omits overt discussions of slavery and the sugar industry from *Voyages*. She depicts instead the ways in which transported Africans (whom she almost always refers to as "Negroes," thereby eliding their status as enslaved people) employ Caribbean plants and animals. Although Riddell (probably in deference to her family) thus fails directly to confront slavery's injustices in her natural history, she does arguably create possibilities for sympathy through cross-cultural identification.[53] Naturalists such as Linnaeus urged scientific travellers to consult different cultures to attain greater knowledge of indigenous botanical and zoological species.[54] In her scientific travel writing, Riddell offers observations of enslaved African communities, providing new information about West Indian resources while illuminating aspects of these peoples' daily lives. For example, she writes that the root of the maniock plant can be "converted into a bread called *cassava:* The Negroes make it their chief diet, and prepare it themselves, which is done by means of a wooden engine, extremely simple in its construction, and managed by a wheel," and she reports that the resulting cakes "when toasted and buttered, are crisp and very pleasant to the taste" (93). Relating Africans' uses of this plant, Riddell instructs Britons about its process of manufacture and introduces a new food to be desired. Further highlighting these communities' resourceful subsistence on natural objects, she recounts that tobacco "is much cultivated by the Negroes here, who are extremely fond of it," that they sometimes use branches of the cabbage tree to construct "cradle[s] for their children," and that they breed goats with great success "for their own profit" (97, 98, 54–55).[55] Riddell thereby presents enslaved people as working to obtain comfort, provision for their families, as well as economic gain—concerns shared and respected by her British readers.[56] Such gestures toward cross-cultural human values, a trope of sensibility, encourage identification and sympathy in her readers, though it does not go so far as to suggest altering the exploitative structure of slave society.

In addition to subtly invoking sensibility, Riddell's attempts to align the lives of enslaved African people more closely with European pursuits, interests, and values may recall earlier ideas espoused by Buffon and other natural philosophers about the possibility of physical, racial transformation in the Americas and West Indies, altering not only Europeans into Africans but also Africans into Europeans. In his 1777 *Histoire Naturelle*, Buffon included the description and portrait of Marie Sabine, born to Black parents in 1736 in New Spain and displaying "light patches on her dark skin."[57] The term "piebald" can now refer to the absence, in certain areas of a human's skin and hair, of cells that produce the pigment melanin. In the eighteenth century, this term was often employed by naturalists to describe horses or other animals that possessed black and white patches, or patches of other colors, as does the multihued parakeet that Sabine holds in the portrait, emphasizing this zoological comparison.[58] As John Wood Sweet asserts, it was Buffon's analysis of Sabine's case that led him to suggest "that this remarkable birth might be due to the degenerative effects of the American climate on African bodies," and "that if there were cases of blacks becoming white, it was only logical to assume that there were whites becoming black."[59] Further supporting this claim, naturalists circulated accounts of children born to Black parents in the West Indies who displayed alterations in skin color, including a "spotted baby girl, named Adelaide, born on a plantation in the West Indies," and John

FIGURE 6. [Non-légendée] ["enfant noir et blanc"], or *Piebald nègre* ("nègres pies"), in Georges-Louis Leclerc, comte de Buffon, *Histoire Naturelle, gènèrale et particulière: Supplèment*, tome 4, plate 2, p. 568 (Paris, 1777), image by De Sève, engraved by C. Guttenberg (Image courtesy of the Linda Hall Library of Science, Engineering, and Technology)

Bobey, who caused a sensation after being brought to London from the West Indies as a young child.[60] These accounts helped substantiate the notion that the West Indian climate and location could facilitate various transformations, racial and otherwise. Moreover, according to Kathleen Wilson, climate theories such as those expressed by Buffon contributed to "eighteenth-century ideas about history involving inevitable progress and degeneration [that] distinguish its evolutionary thinking from its Victorian successors."[61] As Roxann Wheeler notes, "National characteristics that had been considered effects of climate or of differing stages of civilization during the eighteenth century became causes of European superiority and of other races' inferiority by the mid-nineteenth century."[62] In the late eighteenth century, as texts by writers such as Riddell and Buffon demonstrate, the transatlantic was considered a location in which those national characteristics could be in a state of instability, creating the potential for hybridity or transformation.

Nevertheless, even Riddell, despite her sympathies for enslaved people, sometimes portrayed such transformations in zoological terms, like Buffon and numerous other European writers.[63] In *Voyages*, Riddell only once gives an actual voice to enslaved Africans in the British West Indies, when she tests some of the strategies they employed on Antigua while interacting with natural objects on the island. Whereas her text in some ways displays similarities between Europeans and Africans, she here seems to highlight difference. Repeating the Africans' actions, she conducts her own experiment, confirming that the "tradition" describing the island's ground lizard "being attracted by, and fascinated with, the sound of music, is a fact of which, by experience, I can assert the veracity. I have frequently, when sitting in the garden, sung an air in a soft voice, which, in a few minutes, would draw the lizards from the shrubs and trees around to the spot where I was" (65). She explains the origination of this tradition: "It is a well known fact in the West Indies, that, when the Negroes want to catch lizards, (which are a wholesome and favorite food with them) the art they employ to allure them into their hands is whistling" (66). Riddell draws on enslaved people's knowledge of biological species to convey their useful insights into the Caribbean natural world, on which their health and survival depends. Yet, in this single instance of giving Africans a voice rather than merely recording their actions, that voice comes in the form of a whistle. While potentially musical, whistles preclude the use of words and may be interpreted as more primitive, like the calls of island birds and animals, and simultaneously consist of an intriguing unknowableness that masks knowledge, becoming a source of power in itself and mediating between Riddell and the natural environment. Thus, in her travel writing, while Riddell relates her studies, observations, and experiences of less-often-recorded communities in the West Indies, and in some instances provides possibilities for sympathy and identification, she also arguably perpetuates certain racial stereotypes or Eurocentric characterizations within colonial slave society.

Functioning as a naturalist, Riddell both imposes order on the natural world and registers ideas of hybridity that imply transformations often correlated with transatlantic culture. Although *Voyages* does not contain the overt abolitionist sentiments that are present in her personal letters, by sometimes destabilizing differences between the metropole and its transatlantic colonies and disputing the location of degeneration, Riddell helps to alter the era's conceptions of sociobiological possibilities for improvement as well as of capacities for identity transformation. Recognizing such potential for change and multiplicity, she presents alternative ways of conceiving (and embodying) sociobiological implications for gender, sex, race, and nation, as she informs her British audience's perspectives on the knowledge practices of white women colonists, Indigenous peoples, and enslaved Africans.

Metrical Miscellany: Darwin, Sensibility, and Transforming Flowers into Poetry

In Riddell's remaining seventeen years after returning to Britain, she became active in important social and literary circles and published notable texts in addition to *Voyages*, including her memoir of Robert Burns and an edited collection of poetry—two works that, for the purposes of my study, are particularly significant, due to their connections with both sensibility and Erasmus Darwin. After Burns's death in 1796, in testament to her own close relationship with the bard, some of his friends asked Riddell to write a sketch of his life, which she reluctantly agreed to do on the condition that it be published anonymously.[64] Riddell's memoir of Burns has proven invaluable to subsequent biographers and has been praised as "so admirable in tone, and withal so discerning and impartial in understanding, that it remains the best thing written of him by a contemporary critic."[65] Over the course of her friendship with the bard, he wrote many poems for and about her and provided encouragement and suggestions regarding her own verses. Riddell evinced genuine affection and reverence for her friend not only during his life but also through her care in preserving his memory and in resolving his affairs following his death. Five days after Burns's funeral, in another gesture of private grief and devotion, she went by night to plant laurels on his grave.[66] Advocating for a monumental gravestone for Burns, she wrote that it "should be characteristic of him to whom it is raised" and display a "first rate" inscription, hoping that Erasmus Darwin might lend his "talents for the purpose, [as] it could not be given into better hands."[67] This speaks both to Riddell's admiration of Darwin's botanical poetry and to her possible personal correspondence or interactions with this naturalist poet. Indeed, during the late 1790s and the years following the turn of the century, her diary filled with the names of significant personalities of the era whom she "was in the habit of meeting at balls, suppers, parties, receptions, or at the theatre."[68] These impressive figures included the King and Queen, Prince of Wales, Duchess of Gordon, Duchess of Devonshire, "Monk" Lewis, Fox, and Sheridan, among others.

In these years, Riddell also invited her literary connections to contribute to her edited collection of poems, *The Metrical Miscellany* (1802).[69] In 1802, she not only published this collection, which gained enough success to merit a second edition in 1803 and positive reviews in, for instance, the *British Critic*, but this year also saw the publication of the second edition of *Voyages*.[70] Considering its star-studded list of contributors, *The Metrical Miscellany* has received surprisingly little attention from modern scholars. This work "chiefly" boasts "hitherto unpublished" poems from writers such as Riddell; Darwin; Henry Erskine; Anna Barbauld; R. B. Sheridan; Richard Cumberland; Georgiana, Duchess of Devonshire; Charles James Fox; and Henrietta O'Neill.

As the collection's editor, Riddell begins this volume with an introductory poem in which she employs the common botanical metaphor, transforming flowers into poetry. Emphasizing the previously unpublished quality of most of the included works, she describes the flowers/poems as "wild and unshelter'd," so that "the Muse has sought with patient care / 'Mid secret wilds, and meads untried, / A various chaplet to prepare" (ll. 9, 14–16). Thus, Riddell portrays the Muse—or, rather the editor (herself)— as a botanist who searches "near tangled brake, or stream" until "a flow'r we mark, that sure we deem / Is all too fair to blush unknown" (ll. 5, 7–8). In this way, she versifies and aestheticizes the careful process by which she collected these flowers/poems from her contributors and, by wishing for readers who possess "a heart to gentle Friendship prone," sets the tone for attention to both sentiment and the natural world in the ensuing pages (l. 18).

The collection's poems primarily convey themes of sensibility, and several focus on botany or other fields of natural history. For example, the volume incorporates O'Neill's "Ode to the Poppy," Shaw's "On a Butterfly Bursting from its Chrysalis," Riddell's "On a Red-breast," Roscoe's "To a Lily, Flowering by Moonlight," and the Duchess of Devonshire's "The Passage of the Mountain of St. Gothard," which reads as a versified travel narrative with numerous prose footnotes providing details about geographical landmarks, people, and her personal experiences. Darwin's included poems, "The Dream" and "Ode to the River Darwent," prove important components of his oeuvre, especially as they convey insights into his personal life and emotions.

As I especially discuss in the introduction and chapter 6, contemporary critics and writers including John Aikin, Anna Seward, Wordsworth, and Coleridge accused Darwin and his scientific verses of lacking feeling, yet these works in Riddell's *Metrical Miscellany* significantly display his capacity for sensibility. Darwin penned these two poems in 1778, when one of his medical patients, Elizabeth Pole, became ill with a violent fever. Although Pole was married, Darwin fell in love with her and began writing poems to her in 1775, and he composed more than twenty such poems for her over the next six years, which their son, Francis, would later gather together in a collection entitled "A Poetical Courtship."[71] Desmond King-Hele suspects that Darwin's enthusiasm for gardening, which led to *The Loves of the Plants*, may have been inspired by Pole, who was herself an avid gardener.[72] Pole's husband died in November 1780, and Darwin married her a few months later in March 1781, but, at the time of writing these two poems in 1778, his love seemed to have little hope of success. Treating her illness one evening that he thought would prove critical for the disease, he was not invited to spend the night, so he posted himself all night beneath a tree outside of Pole's room in an anxiety-ridden "romantic vigil" recorded in his poem "The Dream":[73]

> Dread Dream! that hovering in the midnight air,
> Clasp'd with thy dusky wings my aching head;
> While to imagination's startled ear,
> Toll'd the slow bell for bright Eliza dead. . . .
> Dream! to Eliza bend thy airy flight,
> Go tell my charmer all my tender fears;
> How love's fond woes alarm the silent night,
> And steep my pillow with unpitied tears. (ll. 1–4, 21–24)

Darwin celebrated Pole's subsequent recovery with another poem, "Ode to the River Darwent," referring to the Derwent in Derbyshire from which his surname derives.[74] A landscape poem of eight quatrains, the ode describes the river's passing scenery while rushing from its source to Derby, where Pole may be found. Here, Darwin exhorts the river:

> Oh! should Eliza press the morning dew,
> And bend her graceful footsteps to your brink. . . .
> Bid your gay nymphs portray, with pencil fine,
> Her angel form upon your silver ground. . . .

> And tell her, Darwent, as you murmur by,
> How in these wilds with hopeless love I burn,
> Teach your lone vales and echoing caves to sigh,
> And mix my briny sorrows in your urn. (ll. 17–20, 23–24, 29–32)

Although Darwin wrote these poems for "Eliza" decades before their inclusion in Riddell's collection, it is tempting to wonder if their publication at this later date may indicate additional personal meaning for him. As I discuss further in chapter 6, his son Erasmus junior committed suicide in 1800 by drowning in the Derwent, and Seward controversially suggested that Darwin felt no grief at this loss. Since "The Dream" expresses sadness and fears for the death of a loved one, and "Ode to the River Darwent" (altered from Derwent to more closely resemble his family name) ends with a line in which the speaker emotionally states that he will "mix [his] briny sorrows" with the river that he calls an "urn," these verses may have resurfaced in Darwin's mind at this time as a representation of his muted, private, continual mourning for the loss of his son—a possibility that further enhances the value of these poems and of Riddell's edited collection. Riddell herself published eighteen poems in *The Metrical Miscellany*, and, in one poem, "Elegy on the Death of Captain J. Woodley," she likewise grieves the drowning death of a family member, her brother, who sank with his ship off the coast of Madeira in 1795.

It is also in *The Metrical Miscellany* that Riddell published her "Inscription Written on an Hermitage in One of the Islands of the West Indies" which, a footnote explains, was composed when she was sixteen. The poem depicts the speaker and her friend Anna in a paradoxical paradise filled with classic British literature and exotic vegetation. The two young women immerse themselves in reading Shakespeare, Spenser, Milton, and Pope while in the midst of the Caribbean's tropical, botanical bounty, so that each morning

> I press with nimble feet the lawn,
> Eager to deck the favorite bow'r
> With every opening bud and flow'r
> Explore each shrub and balmy sweet
> To scatter o'er my mossy seat,
> And teach around in wreaths to stray
> The rich Pomegranate's pliant spray,

> At noon, reclin'd in yonder glade,
> Panting beneath the Tamarind's shade,
> Or where the Palm-tree's nodding head
> Guards from the Sun my verdant bed,
> I quaff, to slake my thirsty soul,
> The Coco's full nectareous bowl. (ll. 31–43)

This glimpse into Riddell's youthful interactions with West Indian botanical species anticipates her later technical, scientific reportage on numerous particular plants in *Voyages*. In the poem's balance between the speaker's studies of British literature and enjoyment of exotic plants, it also highlights Riddell's repeated juxtaposition between her "native realms" of "distant Albion's blissful plains" and "India's spicy bow'rs" (ll. 55, 16, 57). In this way, she expresses consciousness of her own transforming national identity, admitting that she can never "forget" either locale's exertions of influence on her being (l. 56).

While still glancing toward the West Indies, my next chapter focuses on a British colony situated much closer to the metropole—Ireland—and a novel by Sydney Owenson, another author who felt some confliction within her national identity. Like Riddell, Owenson employs knowledge of colonial botany to promote British readers' sympathy or identification with certain aspects of her chosen locale, but Owenson places greater emphasis on botanical analogies when portraying colonized people and thereby highlights the threat of imperial consumption. For Owenson, botany and sensibility thus provide important contexts for her anticolonial literary strategy.

4

Cultivated for Consumption

Botany, Colonial Cannibalism, and National/Natural History in Sydney Owenson's *Wild Irish Girl*

THE 1800 Act of Union has been called "the defining moment of modern Irish history," bringing into effect the United Kingdom of Great Britain and Ireland.[1] Nevertheless, historians such as Patrick Geoghegan note its failings in, for instance, the English government's use of bribery and corruption to secure the act's passage in the Anglo-Irish Parliament while not granting Catholic emancipation, so that this "genuine, if flawed, attempt" at union merely "endured, rather than flourished, after 1801."[2] As Declan Kiberd explains, two perspectives developed, largely divided by the Irish Sea itself, in which the union represented to some "a benign offer of membership in one of the greatest organizations in human history" and, to others, "the most insidious of all oppressive tactics."[3] In the context of this national union, the poet and novelist Sydney Owenson often invoked her own hybrid nationality, resulting from the union of her Irish Catholic father and English Protestant mother, and "made no secret" of preferring her Irish heritage.[4] Scholars generally recognize her national tale *The Wild Irish Girl* (1806) as allegorizing the Act of Union, but there is less agreement about whether her depiction is meant to be politically "defiant" or conciliatory.[5] Although critics have overlooked Owenson's use of sensibility in conjunction with botanical themes and imagery throughout her novel, I argue that she employs this framework to encourage British readers' sympathy for the Irish by subtly highlighting the unsavory aims and effects of England's imperial absorption of Ireland.[6]

Indeed, Owenson begins her national tale with an epigraph taken from an early modern European traveler's reflections on Ireland, "England's first overseas colony."[7] Imputing to the Irish a barbaric appearance amid "rough," mountainous, difficult-to-cultivate soil, the traveler remarks,

> This race of men, tho' savage they may seem
> The country, too, with many a mountain rough,
> Yet are they sweet to him who tries and tastes them.[8]

This rhetoric of cannibalism ("tries and tastes") begs the question of where savagery lies in imperial conquest and sets the novel's tone for conceiving the Irish as plant-like commodities, as nourishing food for the "body politic" through Ireland's incorporation (or ingestion) into Great Britain with the Act of Union. As Creighton Nicholas Brown remarks, "Cannibalism embodies the imperial project" in which "the British Empire consumed distant lands—lands filled with native peoples—to increase its global dominance."[9] Depicting England's seizure of economic and political control over Ireland as a form of cannibalism, Owenson presents additional kinds of English consumption that result, for example, in the loss of Irish national identity and of personal property rights as symbolized through contemporary laws of coverture in marriage, all of which she represents through botany.[10]

Owenson repeatedly describes her English protagonist, Irish heroine, and the Irish population at large in terms of flowers and vegetables, sometimes delineated in great scientific detail, underscoring symbolism and use value, as well as contentions of class and nation. While she identifies the Irish lower classes with potatoes and common herbs, including rosemary, Owenson frequently compares the English and Irish upper classes with cultivated flowers. Within such botanical associations, the English male protagonist cultivates his future wife, the Irish Glorvina, through sentiment and botanical education, just as she cultivates him through education in Irish natural objects, history, language, and tradition. Each stage of their courtship is emblematized through specific plants, as when a rose becomes the chalice or wedding cup that seals the union of these two main characters as well as of their separate nations.

However, this union of England and Ireland conveys not only the potential organic health and aesthetic beauty of national and natural hybridity but also the dangers in such sociological "engrafting." For instance, the novel's Irish-Scottish bard, known as "the mon wi the twa heads," may be likened to a monstrous double flower, suggesting that such national mixture can lead to deformities in both physical bodies and moral character. Additionally, Owenson's comparisons between the Irish peasantry and enslaved Africans in the West Indian colonies exacerbate notions of British imperial cannibalism through botanical imagery. In the context of works such as Erasmus Darwin's *Loves of the Plants* and popular contemporary interest in "the language of flowers," Owenson melds ideas of science and sensibility, as well as natural history and national history. In doing so, she explores Ireland's past flourishing and the possibility, couched in these vegetable analogies, for future cultural regeneration within Irish soil despite the constant threat of English consumption of all that grows there.

Ireland's Colonial Cannibalism

In Owenson's novel, the English male narrator, Horatio, is sent by his father, Lord M., to visit their estates in Ireland for the first time in hopes that this experience will reform the son from his profligate past. While in Ireland, Horatio discovers that his family murdered the ancestor of the Irish Prince of Inismore during the Interregnum in order to attain these estates. Avoiding the familial enmity that plays out national tensions, he therefore introduces himself to the Prince and falls in love with the Prince's daughter, Glorvina, by an assumed name. Horatio's father, an absentee landowner, also courts Glorvina under a false name, and, when both suitors' identities are revealed, the elderly Prince dies of shock (is "murdered") at the prospect of an alliance with his hereditary adversary (242). Nevertheless, Glorvina agrees to marry Horatio in a union that will conjoin the families and, symbolically, their respective nations.

In the wake of the Act of Union, Owenson's national tale largely owed its popularity to providing English readers with better understanding of their Irish neighbors, their new "sister kingdom."[11] As Mary Campbell explains, Owenson herself "supported Catholic emancipation, deplored the Act of Union, the civil disabilities imposed on the Catholics, and the consequent degradation of the native Irish," and therefore sought to improve English perceptions of Ireland as a means to improving Irish conditions within this union.[12] Owenson addresses negative conceptions of Ireland when Horatio confronts his own expectations of the nation. He acknowledges that his view of Ireland derives from travel narratives read in his youth, portraying, "so late as the days of Elizabeth, an Irish chiefton and his family . . . frequently seen seated round their domestic fire in a state of perfect nudity" (13). For him, this conjured up an image "illustrative of the barbarity of the Irish at a period when civilization [in other nations] had made such wonderful progress," so that he admits, "whenever the Irish were mentioned," he could only think of "an *Esquimaux* group circling round the fire which was to dress a dinner, or broil an enemy" (13). Owenson addresses this long-standing stereotype of the Irish as savage cannibals to emphasize its political implications for Anglo-Irish relations.

English writers historically depicted Irish cannibalism to justify inflicting imperial violence on, and absorption of, this "less civilized" country. In *A View of the Present State of Ireland* (1596), Edmund Spenser suggests that, rather than pacifying the barbarous Irish race, the English should implement a scorched earth policy so that the Irish "would quickly consume themselves, and devour one another."[13] Elizabeth Rawdon, Countess of Moira, to whom Owenson dedicated her volume of *Poems* (1801), published an article in *Archaeologia* in 1785 describing the methods of famine the English "continued to employ in Elizabeth's reign, to civilize the Irish."[14] She quotes from Fynes Moryson's *History of Ireland* (1735), which documents eyewitness accounts of these imposed famines, resulting in, for instance, "a horrid spectacle of three children, the eldest above ten years old, all eating, and gnawing with their teeth, the entrails of their dead mother, upon whose flesh they had fed twenty days past," and "some women . . . executed at Newry for killing and eating children."[15] Sympathetically contextualizing this Irish desperation in reaction to the starvation enforced by English policies, the Countess of Moira states, "It seems but candid to seize any opportunity of relating what the antient Irish endured from the English, since" the English typically describe such Irish "cruelties . . . as not having arisen from a provocation."[16] As Owenson makes clear, English "provocation" was elided in the accounts of Irish savagery Horatio read in his youth. Of course, Jonathan Swift's bitter satire *A Modest Proposal* (1729) also famously alludes to Ireland's reputed cannibalism while establishing English responsibility for, and modeling of, such behavior. Condemning the English and Anglo-Irish for compounding Ireland's economic crisis, Swift offers anthropophagy as a solution for overpopulation of the Irish poor and deems English consumers the best prospective customers because they are already accustomed to (commercially and politically) cannibalizing colonial others, especially the Irish.[17]

FIGURE 7. The political satirist James Gillray portrays a horrifying scene of cannibalism in Republican France in his cartoon *Un Petit Souper a la Parisienne;—or—a Family of Sans Culotts, Refreshing after the Fatigues of the Day*, 1792. (© National Portrait Gallery, London)

Owenson's associations of cannibalism with the Act of Union, which formalized Ireland's colonial status and was "achieved by bribing an exclusively Anglo-Irish Parliament, and without the consent of the native Irish," additionally could invoke for readers accounts of recent atrocities, including cannibalism, committed by Parisian crowds during the French Revolution.[18] Edmund Burke's *Reflections on the Revolution in France* (1790) deplores the "cannibal appetites" of the regicidal masses, and, as Denise Gigante relates, "by the end of the eighteenth century, the dichotomy of 'civil' and 'savage' had been mapped onto Europe itself, [in] the Continental extremism associated with French aristocrats and their antithetical starving multitudes," an extremism that parallels Swift's contrast of the Irish "starving multitudes" and the English/Anglo-Irish ruling elite.[19] In fact, ideologies of the French Revolution inspired the Irish Rebellion of 1798, an uprising against British rule in which the United Irishmen were aided by French troops and later defeated, captured, and massacred, along with many Irish civilians, by the British Army, thereby heightening the British government's sense of urgency in bringing about the Act of Union. As Michael de Nie writes, through comparisons between French sansculottes and Irish resistance, depictions of "violence, inhumanity, savagery, madness, and cannibalism . . . would all appear repeatedly in the British press accounts of the Irish rebels of 1798."[20] Voltaire captures England's enduring fears of such colonial uprisings in Ireland in his essay "Anthropophages," in which a woman chandler in Dublin during the Commonwealth era sold "excellent candles made of English fat."[21] Thus, tensions surrounding the Act of Union enable Owenson to evoke these historical, literary, and political legacies in which the Irish both cannibalize and are themselves consumed by their English neighbors.

Moreover, Owenson interestingly places her cannibalistic references within a discourse of colonialism that associates national history with natural history, particularly botany. A hint of this connection appears in the Countess of Moira's accounts of anthropophagy, during English-provoked famines in Ireland, relating that "no spectacle . . . was more frequent in the ditches of Irish towns, and especially in wasted countries, than to see multitudes of poor people dead, with their mouths all coloured green by eating docks [herbs in the *Rumex* family], and all things they could rend from the ground."[22] Such haunting images of hunger and desperate dependence on vegetable nourishment remained connected with Ireland in Owenson's day. In the late eighteenth and early nineteenth centuries, Britons published numerous taxonomies of local flora, and naturalists often drew comparisons between natural objects and national characters so that nature frequently assumed national implications.[23] When Owenson published her novel, the anthropomorphism of plants constituted a popular literary trope, perhaps most famously explored in Erasmus Darwin's long scientific poem *The Loves of the Plants* (1789), which reverses Ovid's "poetic transmut[ation]" of men and women into trees and flowers to "restore some of them to their original" human forms, and his scientific text *Phytologia* (1800), which investigates plants' sentiency and mobility.[24] Such works created a context in which sociobiological analogies carried recognizable precedents and meanings. In this cultural climate of comparisons between vegetation and people, the eating of plants paradoxically could invite metaphors with cannibalistic interpretation.

Cannibalizing Plants: West Indian Slavery and Irish Poor in the "Potatoe Ridge"

By the time Owenson wrote her novel, Britons already were seeing powerful political results from public representations of colonial oppression and cannibalism through botanical imagery. Published in 1806, *The Wild Irish Girl* appeared in the year before Britain abolished the African slave trade. In the two decades leading up to this legislation, a chief tactic of antislavery pamphlets had been to conflate, either figuratively or literally, enslaved African laborers in the West Indian colonies with the sugar cane they produced for British consumption. For example, Andrew Burn's 1792 anti-sugar pamphlet seeks "to convince the inhabitants of Great Britain, who use Soft Sugar, either in Puddings, Pies, Tarts, Tea, or otherwise, that they literally, and most certainly in so doing, eat large quantities of ['the Blood of the Negro']."[25] In this political context, writing a review for the *Athenaeum*, Owenson hyperbolically depicts the conditions experienced by Irish peasantry as more deplorable than those of enslaved Africans, lamenting that Ireland's landlord and tenant system is "worse than the relations of black and white men in the West Indies. Yes, the condition which exists under the sway of the English law, for Paddy, is more inhuman than pagan slavery, feudal ascription to the soil, or the subjugation of the African."[26] As Nini Rodgers explains regarding a slightly earlier historical era, "the seventeenth-century flow of indentured laborers to the Caribbean surfaces in the memory that Cromwell transported Irish men and women to the West Indies as slaves."[27] In her novel, Owenson makes explicit this comparison between the Irish lower classes and enslaved Africans when Horatio refers to the English steward managing his family's Irish estates as possessing "the transmigrated soul of some West Indian planter," who also describes the Irish tenants as subhuman and in need of a "slave-driver" (34, 31).

While enslaved Africans became identified with their cultivation of sugar cane, Owenson associates the Irish peasantry with vegetation grown in Ireland. She portrays the Irish lower classes as struggling with poverty and malnutrition, and their dependence on potatoes offers both physical salvation from hunger and a metaphor for their "consumption" within the English body politic. Horatio's initial interactions with the Irish peasantry change his perspective of them from barbarous cannibals to "warm-hearted people," imparting "Irish hospitality," so that his "prejudices have received some mortal strokes" (16). During his journey, he gains the company of the indigent Irishman Murtoch O'Shaughnassey, who, along with his wife and six children, is starving, unable to afford a plot of land on which to grow potatoes due to the oppressive Anglo-Irish landowning system (24). The two travelers receive a night's food and shelter from an Irish family of simple means and great generosity. Horatio states, "On being admitted into the social circle, I found its central point was a round oaken stool heaped with smoking potatoes," manifesting the centrality of this vegetable as the "national diet" (28). Without means to grow such "grass potatoes," Murtoch explains that his wife is dying "for the want of" nourishment (28).

In eighteenth- and early nineteenth-century English thought, potatoes were often correlated with the Irish working classes as representing their chief subsistence. As Redcliffe Salaman states in *The Social History and Influence of the Potato*, this vegetable had been a staple food in Ireland since the 1640s, whereas in England, even through the late eighteenth century, potatoes generally were fed only to livestock and associated with Catholicism.[28] As Nick Groom explains, the potato "symbolized Ireland, and therefore Catholicism..., and therefore France, the old enemy, where they allegedly had all-potato cookbooks and Marie Antoinette wore potato flowers in her hair—the potato's spreading tuber roots were themselves symbolic of the nefarious spread of Catholicism."[29] This vegetable lent itself to identification with Irish peasantry all the more due to its "disconcertingly human" eyes and skin.[30]

In this vein, the painter and caricaturist Isaac Cruikshank employs potatoes to portray Ireland's danger of being economically and politically devoured in the proposed Act of Union that clearly appears to Ireland's disadvantage in his political satire *Miss Hibernia at John Bulls Family Dinner* (1799). Here, John Bull, representing England and brandishing a knife, sits at the head of his extensive "family," the members of which personify various taxes (e.g., Walter Window Tax, Polly Powder Tax, Hannah House Tax, Simon Soap Tax) waiting to be "fed," while the largest family member, "Isacc Income," shoves the meal of potatoes, symbolizing Irish assets, toward his gaping mouth, and the rest of the family clamor for their share. Meanwhile, Miss Hibernia, signifying Ireland by wearing a dress decorated with Irish harps, politely protests to John Bull, "I fear I must decline all thoughts of the intended Union—your family is so very large and so enormously expensive, that I fear my small dowry will scarcely suffice." Due to associations between potatoes and Ireland's people and culture, the dinner's marks of outward civility contrast

FIGURE 8. Isaac Cruikshank provides a rare London-based critique of the proposed Act of Union, depicting England's economic consumption of Ireland in *Miss Hibernia at John Bulls Family Dinner*, 1799. (Library of Congress, Prints and Photographs Division, LC-DIG-ppmsca-05434)

uneasily with England's symbolic and savagely insatiable consumption of this vegetable (Hibernia's repetition of "fear" seems no accident), representing the colonial conquest/consumption to be made official in the Act of Union. In Owenson's novel, likewise representing the union as a potential marriage, potatoes signify survival on a national level for the Irish poor as their staple "diet," which the English Horatio happily consumes. Mingling derision and reverence, he describes the Prince's "dominions" as "a potatoe ridge," metonymously populating Ireland with this vegetable in place of its people and thus emphasizing his figurative cannibalism (67).

If potatoes symbolize Irish life, edible vegetation also conjures up emotional representations of death or remembrance for the nation's lower classes when, for instance, a daughter places "a sprig of rosemary" on her father's grave (185). Rosemary's role as a culinary seasoning subtly associates food consumption with memorials to the dead, further echoing symbolic cannibalism in the English and Anglo-Irish appropriation of Irish resources, responsible for starvation and deaths among the Irish peasantry. Throughout her text, Owenson not only *likens* the Irish lower classes to plants but also describes them *as* vegetation. She employs plant metaphors for Irish laborers, for example, under English imperialism in Ireland, writing, "The hand of prejudice and illiberality has sown the seeds of calumny and defamation, to choak up those healthful plants, indigenous to the soil, which still raise their oft-crushed heads, struggling for existence, and which, like the palm-tree, rise in proportion to those efforts made to suppress them" (178). Thus, despite English and Anglo-Irish oppression, the Irish peasantry, as "healthful plants, indigenous to the soil," demonstrate proud and enduring resilience.

In addition to conveying Ireland's associations with cannibalism and colonial exploitation, plants act as a means of differentiating between the Irish classes. Glorvina explains to Horatio that, in Ireland, color of "costume" traditionally "became the mark by which the different classes of the people were distinguished," with "bright yellow" serving as the favorite of the lower classes. Glorvina states, "I believe formerly, as now, they communicated this bright yellow tinge with indigenous plants, with which this country abounds" (94). Colored by flowers, this social taxonomy of costume suggests a correspondence with botanical classifications ordering the plants used in the dying process, an artificial system here determined by color as the single character.[31] Employing species' scientific names, she directs him to "see this little blossom, which they call here 'yellow lady's bed-straw,' and which you, as a botanist, will better recognize as the *Galicens borum*; it communicates a beautiful yellow; as does the *Lichen juniperinus*, or 'cypress moss,' which you brought me yesterday; and I think the *resida Luteola*, or 'yellow weed,' surpasses them all" (94).[32] Although Glorvina describes the "inexhaustible" "botanical treasures of our country" as "little known" to the outside world, she declares that "there is not a peasant girl in the neighborhood, but will tell you more on the subject" (94). Presenting the Irish as well informed about their nature and culture, Owenson's botanical specificity is thus for the benefit of her British readers.

A decade and a half earlier, the English naturalist Gilbert White, in his *Natural History of Selborne* (1789), affirmed Owenson's portrayal of Ireland's plants as "little known" beyond its national borders. He hoped that some zoologist would "extend his visits to the kingdom of Ireland; a new field, and a country little known to the naturalist," and that he would be accompanied "by a botanist, because the mountains have scarcely been sufficiently examined; and the southerly counties of so mild an island may possibly afford some plants little to be expected within the British domains."[33] Within the context of Enlightenment determinism, in which climate and soil were thought to shape national identity, White's description of Ireland as climatologically and geographically disparate from "British domains" suggests expectations of difference in Ireland's people as well as its plants.[34] Knowing that it would be more difficult for Britons to rationalize the subjugation of a nation familiar and, in many ways, similar to their own, Owenson aids in the understanding of both the population and the botany of this newly acquired kingdom of Britain, elevating its lower classes as more knowledgeable than naturalists about Ireland's national, natural history. This edifying exchange between Glorvina and Horatio also exemplifies the dynamic of mutual cultivation, integral to their courtship and the symbolic union of their nations.

Sensibility, Education, and Courtship: Cultivating Literary and Botanical Tastes

Although the English Horatio and the Irish Glorvina engage in efforts to cultivate or educate one another as part of their courtship, they do so in different ways. Upon her introduction in the novel, Glorvina appears as the ingenue, well versed in subjects such as history, language, music, and science, but inexperienced in love. Horatio seeks to win Glorvina by awakening her sexual awareness, a progress charted in their botanical interactions. While Glorvina's teacher, Father John, is an "enthusiast" of botany and educates her in its technical aspects so that she "know[s] every leaf that grows" as well as plants' scientific names within the Linnaean system, Owenson emphasizes Glorvina's ignorance of this taxonomy's sexual connotations (58). Sometimes deemed inappropriate in eighteenth-century women's learning, Linnaeus's "sexual system" of botany classified plants according to their reproductive parts, famously inspiring invective from conservatives such as Richard Polwhele, who noted, "Botany has lately become a fashionable amusement with the ladies. But how the study of the sexual system of plants can accord with female modesty, I am not able to comprehend."[35] As I have discussed, regardless of—and doubtless partly because of—this scandalous potential, Linnaeus's sexual system of botany, encouraging the imaginative anthropomorphism of flowers, became extremely popular, as evidenced and furthered by its versification in Darwin's *Loves of the Plants,* where flowers' amorous activities comprise the central theme.

Unlike Glorvina, Horatio demonstrates understanding of botany's sexual implications and sometimes exploits the analogy between women and flowers, as well as the botanist's gaze on flowers' reproductive anatomy, to playful effect. After Glorvina presents him with a violet, he recounts, "I took it in silence, but raised it no higher than my lip—the eye of Glorvina met mine, as my kiss breathed upon her flower: Good God! what an undefinable, what a delicious emotion thrilled through my heart at that moment! ... I gazed at her beautiful flower" (80). Horatio exhibits heightened emotion, erotically associating Glorvina with "her beautiful flower." Glorvina, however, initially lacks such sexual consciousness, as is illustrated when Father John remarks that the rose is the flower of love, and Glorvina "pluck[s] a thistle" and "bl[ows] away its down," botanically dramatizing her disarray of thoughts even as she tries "to hide her confusion" (81). Emphasizing her naivete, Horatio describes Glorvina as "a creature who talks of a violet or a rose with the artless air of infancy, and yet fascinates you in the simple discussion" (83). Glorvina's ability to converse about botany in precise technical terms indicates that her "artless air of infancy" applies not to her scientific acumen but to her ignorance of botany's sensual and even pornographic potential. It is her sexual innocence and scientific mindset that Horatio seeks to dispel.

In her volume of poetry *The Lay of an Irish Harp* (1807), published the year after *The Wild Irish Girl*, Owenson assumes an authorial persona that inverts Glorvina's intellectual relationship to botany.[36] Owenson's notes to her verse display conventional modesty, denying her technical understanding of botany and instead stressing sensibility in her interactions with nature. In her notes to "The Violet," she dismisses scientific acumen, stating, "Without being deeply studied in Linnaeus, or knowing scarcely more of *Bonet, Ludwig,* or *Zunguis,* than the titles of their works, the winter's solitary snowdrop, the spring's early violet, the summer's first rose, and the autumn's last carnation, speak to my heart a language it understands, which Nature dictates, and Science could scarcely improve."[37] Although her awareness of these works suggests more botanical expertise than she admits possessing, she indicates that scientific knowledge of flowers is unnecessary to their appreciation and may even diminish the intuitive communion with Nature inspired in the feeling "heart." Likewise, in her poem "The Sensitive Plant," she draws a comparison between this responsive botanical species and a young woman, Emily, whose depth of sentiment and sensation is noteworthy for surpassing that of the plant. Owenson's prose notes to this poem quote from the naturalist Richard Lobb, who extensively references Erasmus Darwin's studies of vegetable motion and sensation, scientifically grounding this physiological affinity between plants and humans.[38]

Hoping to inspire such a botanical mode of sensibility, Horatio educates or "cultivates" Glorvina by giving her several books reputed for being sentimentally and emotionally charged. Among these books he includes Jean-Jacques Rousseau's *Nouvelle Heloise* (1761), Charles-Albert Demoustier's *Lettres sur la Mythologie* (1786–98), Jacques-Henri Bernardin de Saint-Pierre's *Paul et Virginie* (1788), Johann Wolfgang von Goethe's *Sorrows of Young Werther* (1774), and François-René de Chateaubriand's *Atala* (1801). Horatio makes clear the purpose of these texts: "Let our English novels carry away the prize of morality from the romantic fictions of every other country; but you will find they rarely seize on the imagination through the medium of the heart" (144).[39] Previous to these readings, the Irish Glorvina evinced a general affection for her fellow beings, and an "English" "morality" that operated through prudence and intellect. These novels by writers of other European nations now introduce cosmopolitan sentiments of sexual awareness; as Horatio explains, "her days and nights are devoted to the sentimental sorcery of Rousseau; and the effects of her studies are visible in her eyes. When we meet, their glance sinks beneath the ardor of mine, in soft confusion: her manner is no longer childishly playful, or carelessly indifferent, and sometimes a sigh, scarce breathed, is discovered by the blush which glows on her cheek or the inadvertency of her lip. Does she then begin to feel she has an heart?" (149). It seems no coincidence that several of these chosen writers of sensibility, such as Rousseau, Goethe, Chateaubriand, and Saint-Pierre, also specialize in botanical thought, so that Glorvina's new experiences of sexual awareness fittingly manifest in her botanical interactions.[40] In Rousseau's *Letters on the Elements of Botany, Addressed to a Lady*, his well-known interest in flowers couples with his efforts in cultivating women. Here, Horatio cultivates Glorvina's "taste" through these writers of natural history, these foreign cultures' focus on the heart, seducing her and more closely aligning her with conventionally feminine and sentimental qualities. Owenson's novel thus subtly emphasizes ways in which the study of botany could be integrally connected with concepts of sensibility.

Extending the identification of humans with flowers, every stage of the couple's unfolding relationship is symbolized through botanical imagery. For example, Horatio relates being drawn to Glorvina and her harp music in botanical terms that celebrate Ireland with mixed praise: "The roses of Florida, though the fairest in the universe, and springing from the richest soil, emit no fragrance; while the mountain violet [Glorvina], rearing its timid form from a sterile bed [Ireland], flings on the morning breeze the most delicious perfume [her music]" (60). He also portrays himself as a plant, basking in Glorvina's "light . . . whose nutritive warmth cherishes

[him] into existence" (79). After Horatio's fall from a moldering castle wall that allows him to meet Glorvina and her family, she weeps over the snowdrops that have been stained with his blood (61). Flowers repeatedly stand in for one or both of the lovers, and they develop a "language of flowers" between them, reminiscent of that discussed in Lady Mary Wortley Montagu's *Turkish Embassy Letters* (1763), by which individuals "may quarrel, reproach, or send letters of passion, friendship, or civility, or even of news, without ever inking your fingers," and that became increasingly popular in the Romantic and Victorian eras, assigning each flower a meaning.[41] In Owenson's text, Horatio compares an Irish garden with Eden, imagining Glorvina as Eve and himself as Adam while the couple drinks dew from the same rose to mimic the marriage rite of drinking from the same cup and, thenceforth, understand one another as betrothed (141). Horatio later reveals that he has kept the rose as a charm to protect and remind him of their love. Similarly, he gives Glorvina a small branch of myrtle that, in the language of flowers, typically represents marriage. In her later madness, Glorvina identifies this flower as a "bridal gift from her beloved" (237). When the couple are briefly parted, Horatio maintains the flower comparison, but feels bereft of botany's sensibility, lamenting, "I become again like the plant I tread under my feet; endued with a vegetative existence, but destitute of all sensation, of all feeling" (224). His "vegetative existence" here refers to his lassitude and willingly constrained mobility ("I have not power to move"), yet his ending contradiction ("but") indicates that to be without sensation and feeling is not to be plant-like at all (224). Instead, botany signifies these characters' powerful emotions, longings, and eventual marital union through plants' anthropomorphism and symbolism.

The Naturalist's Gaze

Horatio and Glorvina take turns associating one another with flowers and thus alternately assume the role of the gazing, classifying naturalist. In multiple scenes Glorvina blushes to find herself under the steadfast scrutiny of Horatio's eyes, but in several instances it is Horatio who disintegrates into blushing confusion: "Her eye directed towards me. I know not why, yet I felt confused and gratified by this observation" (69). In such moments, Horatio appears feminized as the scrutinized natural object. Repeatedly, in reaction to this emasculation, Horatio correlates Glorvina with monstrosity. He likens her to a syren, gorgon, witch, *sport of Nature*, and *lusus naturae*. In doing so, Horatio recasts her as a dangerous and unnatural figure, conjuring up traditions of Ireland's savage, cannibalistic past, and as the naturalist's nightmare, an anomaly that challenges set orders and taxonomies. In these moments, he portrays Glorvina—and, by extension, Ireland—as something to be conquered, civilized, and feminized so as to be more easily categorized and possessed.

Horatio is, in fact, the premier naturalist of the novel, claiming the profession of "itinerant artist," intently gazing on or observing and sketching both the landscape and Glorvina as a means to attainment. He explains that he "came to Ireland to *take* views, and *seize* some of the finest features of its landscapes," having "*penetrated* thus far into this remote corner of the province" (56, italics mine). His rhetoric of possession and sexual violation establishes the gaze of the artist/naturalist as a covertly aggressive act. As Natasha Tessone states, "Nineteenth-century England was, indeed, 'the Staring Nation' *par excellence*, one whose sense of national identity was inextricably tied to the gaze at the Other staged by museums and related public shows"—a desire Owenson fulfills through recovery of Ireland's language, history, music, and antiquities in her text and notes.[42] While Glorvina rivals Horatio's gaze by, for instance, sketching his portrait, she then defaces or erases her drawing. By contrast, he displays his likeness of her in which she is "turned into an artifact," captured as a natural object, and includes the quotation "'Twas thus Apelles bask'd in beauty's blaze, / Nor felt the danger of the stedfast gaze" (100).[43] Because he writes the quotation on the same paper that contained her sketches of him, above her inscription of his name, it is not entirely clear to whom we should impute "the danger of the stedfast gaze." But it is clearly Horatio who is aware of the gaze's "danger," to which Glorvina responds with three minutes of petrification before running out of the room.[44]

In this way, Horatio resembles Mary Louise Pratt's naturalist heroes of "anti-conquest." Pratt explains that these heroes often seem impotent or androgynous and are portrayed in infantile or adolescent terms.[45] After Horatio falls from the castle wall, he awakens surrounded by the concerned faces of the Prince, Father John, Glorvina, the nurse, and numerous attendants. It is a kind of rebirth, or, as Horatio later puts it, a "regeneration," in which he assumes a new name, a new identity, and rapidly progresses through an "impotent," "infantile" stage and into something akin to "adolescence" (53, 122). As an outsider to Irish culture, Horatio generally functions in the position of student. He learns the language as well as the culture of the nation in which he finds himself, and his "androgyny" arguably manifests through his emotional nature. Additionally, as Pratt delineates for the naturalist hero of anti-conquest, Horatio's "conspicuous innocence" "acquires meaning in relation to an assumed guilt of conquest, a guilt the naturalist figure eternally tries to escape, and eternally invokes, if only to distance himself from it once again."[46] Horatio (like his father) feels immense guilt for the murderous means by which his ancestors took possession of Irish lands. As a representative of English imperialism, Horatio embodies the longing to find "a way of taking possession without subjugation and violence."[47] His guilt of conquest drives him, not to make reparations of the appropriated land, but to take possession of Irish nature, language, and culture as well, in the appearance of assimilation. Thus, even while the naturalist pose emasculates Horatio, it places him in a position to attain Glorvina, and Ireland, without arousing suspicion or resistance.

Hybridity in Nature and Nation

Although Glorvina, the Prince, and Father John argue that Ireland once was "the most enlightened country in Europe" and revel in the ancient roots of its national history, they also trace that history's subsequent colonization and dissemination to other nations, expressing greater difficulty in now conceiving of Irish culture in its "purity" (107, 174–75). Horatio echoes this notion of "pure" national identity, declaring of Glorvina, "I long to study the pure national, natural character of an Irish woman ... as Nature has originally formed her," for "hitherto I have only met servile copies, sketched by the finger of art, and finished off by the polished touch of fashion" (56). As Kathryn Kirkpatrick explains, modern "scholars have criticized this kind of essentialist thinking about national identity" in the novel even as the construction of such "identities depends on identifying insiders in relation to outsiders," yet Owenson also "dramatically undermines the essentialist definition of Irishness her novel appears to endorse."[48] In a dynamic that indicates Ireland's susceptibility to further colonization, Glorvina, as the novel's embodiment of Ireland, expresses a fateful indecisiveness in maintaining this sense of national identity. Horatio states, "She has unconsciously imbibed many of her father's prejudices respecting antiquity of descent and nobility of birth. She will frequently say, 'O! such a one is a true Milesian!'—or, 'he is a descendant of the English Irish,'—or, 'they are new people—we hear nothing of them till the wars of Cromwell,' and so on. Yet at other times, when reason lords it over prejudice, she will laugh at that weakness in others, she sometimes betrays in herself" (117). Through

Glorvina, we see Ireland educated to "laugh at" maintaining its heritage in the face of British imperialism, supposedly as a means of embracing reason and selective cosmopolitanism. Indeed, Horatio is overjoyed to learn that Glorvina, while Catholic, does not follow many of the practices of Irish Catholicism and is "not devoted to its errors, or influenced by its superstitions" (187). As Linda Colley argues, eighteenth-century Britons largely defined themselves through Protestantism, in contrast with the Catholicism of continental Europe.[49] From the English Protestant perspective, Catholic belief in transubstantiation converted the Eucharist into a cannibalistic ritual, consistent with Ireland's barbaric past.[50] However, Owenson arguably reverses these religious associations with cannibalism so that it is instead the English Protestant faith that threatens to devour or absorb Catholicism. According to Thomas Tracy, the novel suggests that, "with the assimilation of the English and Irish cultures, Roman Catholicism will be superseded peacefully, in the natural course of events."[51] The "assimilation" of these two cultures, then, is not envisioned as an equal union that preserves differences, but one in which Irish culture and religion "will be superseded peacefully" over time. Thus, Glorvina's occasional dismissals of traditional Irish culture both help to separate her from conceptions of Irish/Catholic cannibalism and simultaneously exhibit her culture's vulnerability to colonization.

Owenson depicts concern for the continuance of Irish heritage as a legitimate source of anxiety, exemplified in botanical associations with the deformed Irish bard who symbolizes potential dangers of hybrid nationality.[52] Horatio and Father John journey to visit "one of the last of the race of Irish bards" in northern Ireland, which "may in some respects be considered as a Scottish colony; and in fact, Scotch dialect, Scotch manners, Scotch modes, and the Scotch character almost universally prevail" (199, 197–98). According to the Prince, through their pursuance of industry and capital, "on the heart ['the Northerners'] make little claims" (198). Referencing this region's national mixture, he observes in botanical terms that northern Ireland reveals "the Scotch character *engrafted* upon" that of the Irish, creating an "*exotic* branch," and it is here that the old Irish bard resides with his family in a small, impoverished hut (173). Known as the "mon wi the twa heads," the bard has "an immense wen," or growth on the skin, "on the back of his head" (199). In the context of the Prince's botanical descriptions of this "engrafted," hybrid, or "exotic branch," melding Irish with Scottish culture, the bard's deformity also conjures up ideas of double flowers or double blooms: beautiful but weakened luxuriants produced by florists and often disparaged by eighteenth- and early nineteenth-century botanists for their impurities. According to the contemporary botanist William Withering, "Double flowers are monsters" due to their artificial hybridity.[53] As Sam George relates, "In the wild flower or natural species, which represented health and vitality, an unmodified, essential character could be observed. In comparison, the luxuriant was a gaudy, sickly product of society whose true lineage was disguised."[54] The bard's appearance of diseased, monstrous hybridity seems to germinate from his now-confused "lineage," within this cross-pollinating, so to speak, of Irish and Scottish nations.

Authenticating the existence of this "mon wi the twa heads," Owenson includes a long, personal letter from a friend and correspondent in her footnotes, detailing the bard's history and confirming firsthand observation: "Since I saw him last, which was in 1787, the wen on the back of his head is greatly increased; it is now hanging over his neck and shoulders, nearly as large as his head" (201). Such factual footnotes mark a convention of contemporary travel accounts. As Heather Braun explains, "Orientalist literature typically includes extensive footnotes to complement its aims to instruct about foreign worlds," and Owenson's notes educate her audience in Irish culture, appropriating the orientalism imputed to Ireland by English writers.[55] Ina Ferris expounds that, by adopting this characteristic of travel accounts, Owenson "sets up an elaborate subtext of footnotes in which a personal, authorial voice criticizes, revises, commends, and otherwise engages a plethora of texts on Ireland written from different points of view (and sometimes in different languages)."[56] Further, these footnotes echo formal practices within the scientific poetry of Erasmus Darwin, Charlotte Smith, and other contemporaries, in which imagination guides the text, and footnotes lend scientific and historical explanation. The inclusion of footnotes, typical of various kinds of natural history writing, thereby represented a tactic familiar to Owenson's readers, and these informative scientific genres of travel and imagination provide templates for Owenson's own formal choices. Moreover, her lengthy footnotes sometimes, as in her account of the Irish bard, consume entire pages, leaving little or no room for the tale, thus visibly colonizing or cannibalizing sections of the text.

The bard's hybrid monstrosity amid intermixed Scottish and Irish culture arguably cautions against an overly optimistic view of Ireland's union with England. The novel questions the merits of English national character through Horatio's debauchery, deception, and oversentimentality, as well as his father's untruthfulness, which ultimately speeds the

Prince's death, and the English steward's mistreatment of Irish laborers. What here indicates that a union of Irish national character with that of England will have less monstrous results than that which produced the mutant bard? While one might propose that Horatio and Glorvina each possess qualities that Owenson assigns to the other's nation and thus constitute pre-union "Britons," anticipating a "natural" partnership, this would ignore the union's inequality. After her father's death, Glorvina is faced with either refusing the marriage to Horatio and losing all power and property or accepting this union as a subjugated party under the laws of coverture through which women's legal rights became subsumed (or consumed) by those of her husband, with only subordinate potential for exerting influence over her nation's fate. Additionally, in late eighteenth-century botanical theory, many naturalists understood hybridity as an aberration, often producing barren unions (in the tradition of both zoological and botanical "mules" or hybrids) because departing from original forms.[57] According to the Prince and Father John, from its Greek ancestry the Irish national character has "degenerated" only in the intervention and intermixing of other cultures; as such, the novel's correlation between nation and nature forecasts the result of this union as ambivalent at best (176–79).[58]

Colonial Consumptions, Botanical Conclusions

Owenson's book concludes with a letter from Lord M., adopting botanical language to endorse his son's marriage to Glorvina and advise him on how to govern the Irish peasantry after inheriting his family's Irish estates. Lord M. describes Glorvina as a "lovely blossom from the desart where she bloomed unseen," and, although their impending nuptials "blasted" his "parental ambition . . . in the bloom," he stresses the national importance of Horatio's marriage, encouraging it as a means by which "the distinctions of English and Irish, of protestant and catholic, [may be] for ever buried. . . . [with] this family alliance being prophetically typical of a national unity of interests and affections between those . . . who are naturally allied" (247, 249, 250). Despite this rhetoric of national equality, within the novel's colonial perspective, this "bur[ying]" of national "distinctions" indicates the replacement of Irish customs with those of England. Culminating the tale's botanical theme, Lord M. likens the Irish lower classes to plants, to "creatures of the soil," admonishing Horatio to cherish by kindness into renovating life those national virtues, which, though so often blighted in the full luxuriance of their vigorous blow by the fatality of circumstances, have still been ever found vital at the root, which only want the nutritive beam of encouragement, the genial glow of confiding affection, and the refreshing dew of tender commiseration, to restore them to their pristine bloom and vigour: place the standard of support within their sphere; and like the tender vine, which has been suffered by neglect to waste its treasures on the sterile earth, you will behold them naturally turning and gratefully twining round the fostering stem, which rescues them from a cheerless and groveling destiny. (251)[59]

Having cultivated the mind and affections of Glorvina, Horatio now is charged with a similar husbandry toward the botanically analogized Irish subjects, who require, according to Lord M., "the light of instruction" to "[dispel] the gloom of ignorance and prejudice from their neglected minds" so that they might grow under the (English) light of "reason and humanity" that "genially warms and gratefully cheers the whole order of universal nature" (251, 252).[60]

Since the novel extensively details Ireland's rich intellectual and cultural traditions, its closing repetition of the stereotype that Irish "ignorance" and barbarity must be enlightened through English colonization, historically represented in the Act of Union, seems strategically disingenuous. As Joseph Lew reveals, "Toward the end of her life Owenson's reputation diminished partially because of the unpopularity of her anti-imperialistic ideology in Victorian England"—an ideology subtly conveyed in this novel's ending.[61] Portraying the Irish as plants, as "creatures of the soil" that must be "tender[ly] . . . foster[ed]" in Ireland's "sterile earth," Lord M. ultimately rephrases the epigraph's proclamation that "This race of men, tho' savage they may seem," on "many a mountain rough, / Yet are they sweet to him who tries and tastes them." He continues the historical confiscation of Irish lands, enforcing this "cheerless and groveling destiny" of dependence through an agenda for "rescue" that merely cultivates the vegetable-Irish for consumption within the British Empire's body politic. Employing such bleak botanical metaphors of union, it is no wonder Owenson's manuscript first prompted rejection from her publisher, Sir Richard Phillips, who happened to be a vegetarian and feared that the novel "too strongly opposed . . . the English interest in Ireland."[62]

While Owenson's novel employs botany and sensibility to expose British imperial oppression in Ireland, the next chapter brings my study back to local concerns within England through Charlotte Smith's poetry. In her works, Smith, too, sometimes highlights the problems of empire, but here I primarily focus on her conception of the poet-naturalist of sensibility who meaningfully sympathizes and identifies with closely observed botanical species. In this way, she additionally challenges not only scientific taxonomies and naturalists but also the poetic methods of contemporary male poets.

PART III

In/effability

Sensibilities of Description, Classification, and Defiance

5

"On the Green Margin"

Place, Sensibility, and Originality in Charlotte Smith's "Flora"

In *Conversations Introducing Poetry: Chiefly on Subjects of Natural History* (1804), Charlotte Smith portrays the sea as a space of liminal natural objects lacking straightforward or exact categorizations within naturalists' taxonomic orders. Her poem "Studies by the Sea" presents the ocean as a force that "tears down . . . bounds" and displays "innumerous changes," harboring "endless swarms of creatures" in "unfathom'd waves."[1] In the final poem of *Conversations*, "Flora," she more closely explores the sea's borderlands and border-lives, such as zoophytes, that straddle different kingdoms of the natural world, and "plants of the class cryptogamia"— a reference to vegetation in which the reproductive organs are not easily visible, making classification through usual Linnaean methods difficult, if not impossible.[2] In this way, the sea represents one of several locales in which Smith reveals layers of cultural meaning and taxonomic shortcomings with sociobiological implications for gender, class, and nation that may be poignantly associated with a particular "place." Smith describes "Flora" as a revision of Erasmus Darwin's *Loves of the Plants* (1789), which she admires and defends in its propriety; however, while Darwin's poem versifies Linnaeus's sexual system of botanical classification, she instead places less emphasis on plant reproduction or sexuality while also questioning science's taxonomic impulses.

Juxtaposing different ideas of place, Smith's "Flora" traces a trajectory from the literary tradition of the classifiable and playfully feminized garden into "wild uncultured" and, finally, "unknown" scenes, thereby moving away from the location, form, and ideology set forth in Darwin's ostensibly more scientifically assured poem, undergirded by informative notes (ll. 127, 178). In addition to unsettling and reworking Darwin's approaches to botanical species, Smith here parodies the conventional placements of contemporary (male) poets who, in her portrayal, stand at lofty heights and channel all their mental powers to squint into an abstract distance, too absorbed in some remote or indiscernible idea to notice the intricacies of their natural environment. Referencing specific places associated with her personal history, she depicts such masculine oversights as often having disastrous consequences for femininely portrayed natural objects.

In the poem's penultimate stanza, Smith examines how particular locations can highlight the struggle of taxonomic systems to place ambiguity. "From [the] depths" of the sea, knowledge undergoes crisis, stymied by uncertainties (l. 179). Smith thereby exposes liminalities that undermine taxonomic authority and alter her own claims to scientific knowledge as well. Ultimately, in the poem, she abandons contemporary classificatory systems and retains only natural history's reliance on observation. She revels in the seemingly displaced and unnamed, that which can only be communicated through "description," while also questioning the ability of even this descriptive method to fully convey biological organisms (l. 192n). Aligning the poet-naturalist with sensibility, Smith's sea exploration celebrates imaginative possibilities within the natural world and empathizes with species in the absence of strict placement within taxonomic orders, retrospectively destabilizing the poem's early classifications and engagements with scientific and poetic conventions. Through both her interrogations of male naturalists' assertions and her efforts to relate feminine forms of political power and poetic creativity, Smith's emphasis on sensibility and place suggests the potential for literary originality by revealing botanical mysteries within specific locations of the natural environment.

Framing "Flora": Smith's Personal Politics of Place

Although the majority of Smith's "Flora" maps shifts in physical locations and records the vegetable species present within these different environments, the poem's first and final stanzas assume a more philosophical tone regarding plants' effects on humanity's physical and emotional experiences. In the initial stanza, for instance, Smith gestures toward her era's method of justifying young women's botanical studies as uplifting the spirits of "the wan maid" through physical exercise as well as pious and wholesome mental stimulation, so that botany represents a healthy pursuit (l. 19). As various scholars have discussed, in the late eighteenth and early nineteenth centuries the knowledge of flowers was often presented as the knowledge of women.[3] Women were generally encouraged to study flowers, and poetic and scientific comparisons between flowers and women reinforced the tacit assumption that understanding of one brings understanding of "the other." In "Flora," first published in *Conversations* and reissued in the posthumous *Beachy Head, Fables, and Other Poems* (1807), Smith reconsiders this relationship to form a question: If the knowledge of flowers is the knowledge of women, what follows when botanical conventions and taxonomies are shaken? While the poem's first half appears to uphold scientific knowledge, its final stanzas incite skepticism regarding contemporary classificatory systems and the extent to which natural objects can be ordered or known.

Writers of sensibility, including Smith, Jean-Jacques Rousseau, and William Cowper, sometimes presented the study of botany as a welcome escape or distraction from feelings of melancholia and the world's injustices.[4] In a note to her poem's initial section, Smith cites Cowper's *The Task* as confirming that "Flora banishes" feelings of "sadness," "spleen," and "grief" (ll. 460, 457, 455, 459). Indeed, in her opening lines, she suggests that botany and imagination may help humanity not only to forget "the crimes and follies of mankind" but also to "lose awhile the miseries they mourn" (ll. 2, 6). As many of Smith's works indicate, she may have harbored personal motives for highlighting disruptions within sociobiological notions of fixed order. Although born into the gentry, marriage left her at the mercy of her husband's debts, his creditors, and even his verbal and physical abuse; the resulting poverty and marital separation caused Smith to feel *dis*placed from her rightful station in the upper echelons of society.[5] Her father-in-law's will, which could have relieved the destitution faced by Smith and her children, was famously stalled in court for numerous years and remained unresolved until after her death. Smith's disillusionment with the patriarchal systems of marriage, law, and education help to inform her dismantling of masculinist scientific taxonomies in "Flora." While her intersecting inquiries into knowledge, science, and society are arguably most apparent in "Beachy Head," "Flora" offers an exposé of epistemological limits, and is more blatant in its challenges to classification, expressing her poetry's frequent themes of sensibility, sympathy, suffering, and desire for relief. Through botany's "sweet oblivion," then, one may momentarily "lose," or forget, the memory of what's been lost (ll. 6, 7). In addition to the well-known financial woes faced by Smith and her children, and lamented by the poet in the prefaces of some editions of *Elegiac Sonnets*, Smith also had recently lost her beloved daughter Augusta, who died in 1795 at the age of twenty, a bereavement to which Smith alludes in her moving botanical poem, "Reflections on Some Drawings of Plants," and from which her biographer acknowledges that Smith never recovered.[6] Moreover, Smith wrote "Flora" while enduring various bodily ailments, including dropsy, pleurisy, "an accelerated heartbeat," and perhaps uterine cancer, creating physical as well as emotional pain and likely increasing the appeal of botany as a healing, intellectual distraction.[7]

Smith frequently anchors her verse in her personal life, and, in the opening stanza of "Flora," the place to which she allows her "o'erwearied mind" to escape is "the green margin of my native Wey," referring to the River Wey, which had been viewed by "mine infant eyes" in "life's happier Spring" (ll. 1, 11, 12, 10). The River Wey is a tributary of the Thames that flows into West Sussex. In the prose narrative of *Conversations* that surrounds this poem, the mother-teacher, Mrs. Talbot, looks back on her childhood in Sussex as the poem's inspiration. Yet she states that this location of the poem's focus now exists only in her mind, for "the houses which formed the village are now pulled down, and were I to pass through it, I should see no likeness remaining of the place, still represented by a thousand minute circumstances to my memory," thus imparting a somber tone to the poem's framework.[8] In her youth, Smith herself had lived at Bignor Park in Sussex, where she "was free to explore the Downs and seashore, and to play beside—and in—the tidal reaches of the Arun" River.[9] These early experiences of Sussex landscapes and seascapes helped to develop her "interest in botany that found scope in the great local variety of plant life," and these vegetable-rich environments frequently appear in Smith's poetry, as they do in "Flora."[10]

Smith explores botany in relation to notions of place in many of her poems. She presents herself as a new kind of "moralizing" poet-botanist of sensibility, as in her poem "A Walk in the Shrubbery," distinguishing her work from that of previous generations of poets who, she suggests, "have never been botanists" in the sense of possessing accurate and technical scientific knowledge of plants (303, 242). Since individual plant species are generally thought of as being rooted in place and thriving in specific environments, they often lend themselves to depictions of particular locales. For example, Smith's poem "Verses Supposed to Have Been Written in the New Forest, in Early Spring" contains a note where she explains that a species of blue "Wood Anemone (*Anemone Nemerosa*)" can be "found in Whichwood Forest, near Cornbury quarry" and emphasizes the novelty and particularity of this information: "I do not mention this by way of exhibiting botanical knowledge (so easy to possess in appearance) but because I never saw the Blue Anemone wild in any other place, and it is a flower of singular beauty and elegance" (107). In her educational books, she strives to help young readers "learn the names of common plants in their vicinity," as is the goal in her poem "The Kalendar of Flora," first published in *Minor Morals* (1800), tracing the environments and chronology of species of flowers as they appear throughout the year (190).

Yet Smith also sometimes acknowledges plants' ability to defy temporal and spatial limitations. For instance, in her poem "To a Geranium Which Flowered during Winter," she portrays this flower as being transported from Africa to a British "conservatory" or greenhouse where it eludes the seasons and blooms "amidst the wintry gloom" (208–9). In fact, while John Aikin suggested in his *Essay on the Application of Natural History to Poetry* (1777) the subject of bird migration as ideally suited to provide novelty within poetry since it was so little understood by naturalists themselves, Smith takes the topic of migrating species to even greater extremes of originality. In addition to versifying contemporary ideas about birds' seasonal disappearances, she conjectures about the migrations of fish and of plants (291). In a note to "Studies by the Sea," she explains that "Gulph currents are supposed to throw the remains of fruits of tropical regions on the most northern coast of America, and it is asserted that the same fruits are also found on the coast of Norway" (294). Erasmus Darwin, too, wrote about "itinerant vegetables" such as the *Conserva aegagropila*, "found loose in many lakes in a globular form" that "adheres to nothing, but rolls from one part of the lake to another," and the *Fucus natans*, which "floats on the sea in very extensive masses, and may be said to be a plant of passage, as it is wafted by the winds from one shore to another."[11] Nevertheless, despite some plants' capacities to relocate across various distances, Smith's "Flora" instead focuses on botanical species as more localized and as a means of thinking about the contexts of their particular environments.

Significantly, by beginning her poem "on the green margin" of the Wey, Smith sets up the verses' recurring engagement with liminality, for while "margin" can refer to a riverbank or shore, it contains further connotations meaning an edge, border, or boundary, including, interestingly, the botanical boundary line of an organic structure or part of a plant, and the spatial or temporal point of transition between states or epochs, marking a moment in time when some change or occurrence is imminent—as well as, of course, the space on a page between its extreme edge and the main body of written or printed matter, sometimes containing notes and references.[12] Throughout "Flora," Smith explores these margins or borders in addition to the potential placement of herself and other biological species in relation to them. Calling on "Fancy," the speaker asks that this "Queen of ideal pleasure" might "teach me to describe" with "thy magic pencil" the botanical species and "enchanting Goddess of the flowery tribe" (ll. 7–8, 13, 9, 14).[13] This initial mention of description foreshadows the concept's growing importance within the poem. Aligning her verse with the genre of descriptive poetry, Smith appropriates the necessity of description within natural history writings that support the creation of classifications or taxonomies of biological forms. Indeed, Smith increasingly interrogates the relationship between description and classification over the course of "Flora." However, before questioning these scientific ideas, she first presents them in a confidently stable and playful environment within the poem's early stanzas, where Smith's reference to the "enchanting Goddess" conjures up not only Flora as the Roman goddess of flowers but also Erasmus Darwin's Goddess of Botany and the second half of *The Botanic Garden*, *The Loves of the Plants*; Smith thus sets the scene for her exploration of botany along the riverbank to, in many ways, represent the familiar landscape and literary tropes of the garden (l. 14).[14]

Humor, Poetry, and Science: Parodies of the Garden

Within the poem's "vision" of the riverbank, even Smith's "wan maid" assumes a flower-like "fresher bloom" through her attention to botany, drawing for the reader a recognizable parallel with Darwin's verses in which plants become analogous with people (ll. 19, 20). There, Darwin refashions Ovid's "art poetic [that] transmute[d] Men, Women, and even Gods and Goddesses, into Trees and Flowers," by seeking now "to restore some of them to their original animality."[15] Moreover, Darwin's anthropomorphic plants poetically illuminate a fantasy world of plant sexuality supported by his explanatory scientific prose notes. Although Darwin was, of course, not the first to write botanical verse with scientific notes, his poem's popularity influenced some Romantic-era authors in their use of this formal structure.[16] Smith began adding endnotes about natural history to her *Elegiac Sonnets* in 1786, and thus several years before Darwin's poetic publications, but she does subsequently quote or reference his *Botanic Garden* on numerous occasions, and, in a letter to her publishers on August 27, 1799, she praises it as "one of my favorite books."[17] While species' common names are typically richer in poetic association, Darwin employs the scientific names of plants and insects. Smith, on the other hand, wavers between precision and poetics, alternating between the use of Linnaean binomial nomenclature and common names, while nearly always supplying a note for the missing appellation(s). Additionally, in her notes, Smith cites both scientific authorities and poets, such as William Cowper and James Thomson, blurring the line between fact and fancy.

Like Darwin's "Goddess of Botany," Smith's imagined goddess "Flora" reigns over the botanical world and descends "in her leafy car" while "call[ing] the buds to birth" (ll. 24, 26). Employing wit and humor in the style of Alexander Pope's mock-heroic poem *The Rape of the Lock* (1717), Smith satirizes the epic trope of recording the lineage of a hero's weapon, armor, or mode of transportation. In this early section of the poem, she focuses on the trees and plants that comprise Flora's carriage and dress, capitalizing on interchanges of species' common and scientific names that enhance her comedic effect. For example, detailing Flora's carriage, the speaker explains, "Saxifrage, that snowy flowers emboss, / Supplied the seat," and a note reveals that this plant is "commonly called Ladies' cushion" (ll. 33–34). In another instance, the speaker envisions "Scandix" as holding in place the "wandering tresses" of Flora's "radiant hair," and a note elucidates that this plant is also known as "Venus's comb" (ll. 43, 45). Although Darwin also uses comical techniques, women writers of this era sometimes employ wit or humor in their scientific poetry to help defuse perceived intellectual threats or social transgressions that may be assigned to their technical knowledge of natural history.[18] Such poetic practices allow women to skillfully engage with these sciences while maintaining an unassuming sense of distance through laughter about their subject. Of course, this ability to create comedic effect requires understanding of, and facility with, scientific terminology and concepts to successfully exploit their jocular possibilities. This brand of poetic humor thus gestures toward women writers' simultaneous inclusion and exclusion, prudently and productively obscuring their serious participation in scientific methods and discourse. That being said, in this instance, I would argue that Smith here also subtly parodies or critiques Darwin's use of humor in his scientific poetry as creating emotional detachment from the natural objects that she more sensitively appreciates in other sections of her poem where she places greater focus on feeling and sensibility.

In this part of her poem, more directly invoking Pope's *Rape of the Lock*, Smith additionally appropriates for her portrayal of the riverbank the "sylphs" that comprise his mock epic's supernatural machinery, and that Darwin also employs in his *Botanic Garden*.[19] In Pope's work, the sylphs represent the comical, airy souls of past women who now protect young women's chastity, for "what guards the purity of melting maids, / In courtly balls, and midnight masquerades. . . . / 'Tis but their Sylph, the wise celestials know, / Though *Honor* is the word with men below."[20] Part of the humor, then, is in recognizing that the sylphs could be said to make both everything and nothing happen.[21] Relocating these sylphs into the garden, Smith reimagines their purpose, depicting them as gardeners with the task of cultivating and defending "soft buds" and "infant flowers" (l. 52). Significantly, these sylphs allow Smith to engage with two of the most prominent sociobiological metaphors employed by contemporary naturalists in regard to botany.

On the one hand, Smith's sylphs embody imperial, militaristic themes by protecting "vegetable life" from "the Insect race" (l. 58). As "warriors," these sylphs, "for conquest arm'd," battle the spider, aphis, earwig, dragonfly, snail, ant, and beetle, and shield plants from these invading foes (l. 61). Placing these notions of military order and conquest in the space of the riverbank or garden, Smith draws on William Withering's martial analogies in *A Botanical Arrangement of all Vegetables Naturally Growing in Great Britain* (1776). In his "anglicized version" of Linnaeus, Withering "mask[s] any reference to the organs of generation and disguise[s] the sexual character of Linnaean classes and orders."[22] Avoiding botany's sensual associations, Withering instead highlights its combative correlations in which, for instance, Linnaean classification compares "a class to an army; an order to a regiment; a genus to a company; and a species to a soldier."[23] As Sam George writes of Smith's poem, "The influence of Withering is everywhere . . . from the emphasis on native plants to the lack of concern with the sexualized parts of the flower."[24]

On the other hand, the very actions of Smith's sylphs in this militarized, garden-like space arguably allude to plants' sexualization. In Smith's portrayal, part of the sylphs' duties while cultivating these plants includes working to "save the Pollen from dispersing wind" (l. 54). This brief emphasis on preventing pollen's dispersal gestures toward contemporary concerns about the potential for sexual anarchy in Linnaeus's depiction of the plant kingdom. For instance, to revisit ideas from Smellie's *Philosophy of Natural History* (1790), explored in chapter 3, that naturalist derides Linnaean analogies in the "sexual system" between plants and animals or humans, anxiously imagining the hybridizing "consequences" that such sexual reproduction would imply for the vegetable orders. He argues that, in plants, pollen, "by flying promiscuously abroad," could "impregnate different species which happened then to be in a fit condition for the reception of male influence" and thereby "introduce universal anarchy among the vegetable tribes" so that, in a few years, "the earth would be covered with monstrous productions, which no botanist could either recognize or unravel."[25] Smellie contends that, contrary to the "anarchy" of Linnaeus's sexual system of botany, "all laws of Nature are fixed, steady, and uniform."[26] Yet his fears regarding the possibility of vegetable sexuality center on the (female) plant's indiscriminate receptivity to (male) pollen. Thus, Smith alludes to this controversy by employing the sylphs to police plant sexuality and "promiscu[ity]," conjoining Withering's militarized conception of the garden with Darwin's sexualized portrayal in *The Loves of the Plants*. Additionally, in Smith's focus on the anatomy, the bodily, material forms of plants, when, for example, she names Flora's attendants as Floscella, Petalla, Nectarynia, and Calyxa, after the parts of flowers, she keeps the feminized, sexualized conventions of the garden in view even as she claims to be writing a more "virtuous" version of Darwin's poem (ll. 81, 87, 91, 97).[27]

Smith also arguably alludes to the Linnaean sexual system of botany in the poem's final stanza, where she reflects on the locations explored in her text and delineates which admirers of flora may be drawn to these spaces. While she designates the garden/riverbank and mead, or meadow, as places for children and, arguably botanophiles, of a "gay heart, and frolic step," she additionally associates with that early section the ideas of innocence, beauty, and marriage. However, although Smith here ostensibly advocates marriage as a "bless'd" state, she also provides the possibility of a darker reading. The speaker encourages Fancy to "Bind the fair wreath on Virgin Beauty's brow, / And still may Fancy's brightest flowers be wove / Round the gold chains of Hymeneal love" (ll. 212–14). Smith's language undercuts the lines' seeming endorsement of marriage. The binding of a young woman's brow could signify the limitations of her education, rights, or choices in regard to marital expectations. Moreover, these flowers are of "Fancy" and thus partake of the unreal and even delusive realm of imagination, and are described as the "brightest," perhaps alluding to the florist's cultivated, artificial, monstrous varieties, denigrated by botanists.[28] In Smith's portrayal, the marriage institution is "b[ou]nd" by "gold chains," conjuring up the image of a gilded cage. This "chain" also, in the context of the "great chain of being" which is extolled, for instance, in Pope's *Essay on Man* (1733–34), contains connotations of hierarchies within orders and systems, as well as within gender and social classes.[29] As Linnaeus's sexual system depicts parts of flowers as "husbands" and "wives" within a "marriage," these lines additionally may subtly indict such botanical classifications as reinforcing inequalities through marital, sexual, and social hierarchies. Privileging a plant's male parts (stamens) as determining its class over its female parts (pistils), which signify its order, Linnaeus's botanical system thus imposed traditional notions of gender hierarchy.[30] Considering Smith's experience of personal abuses within marriage, it is perhaps unsurprising that she implies that its potential for happiness exists only in "Fancy."

Balancing fancy and facts within this familiar space of the garden/riverbank, Smith's prose notes to her poem confidently assert scientific knowledge about plants and insects, as well as provide commentary about their aesthetic appeal. In addition to drawing on the writings of other poets and naturalists, Smith demonstrates her own close botanical observations, as in her note to *Tradescantia* where she remarks of the "silk-like tuft" within this plant that, "on examining one of these small silky threads through a microscope, it looks like a string of amethysts." As Judith Pascoe states, Smith "seems to arrive at the magical by way of the empirical.... Scientific observation, rather than negating the musings of the imagination, seems for Smith to initiate them."[31] Nevertheless, Smith also reveals the dangers and responsibilities for poet-naturalists who assert science's imaginative possibilities when she depicts dragonflies as a species of insect assailants against which the sylphs must defend, yet admits in a note that dragonflies' "introduction here is a poetical license, as [they] do not feed on or injure flowers" (l. 75n).[32] Smith may have felt compelled to include such a caveat regarding her "poetical license" in response to John Aikin's imperatives in his *Essay on the Application of Natural History to Poetry* (1777), a text that she references in her poem "Beachy Head." In his *Essay*, Aikin critiques poets who depart from scientific facts and direct observations in their verse descriptions of natural objects. According to him, "Nothing can be really beautiful which has not truth for its basis," and "where the professed intention of the poet is the description of natural objects, it cannot be doubted that every fabulous idea should be religiously avoided."[33] For Aikin, this close and faithful observation of the natural world provides the means to poetic originality, a theory that influenced Smith's poetry and her concerns with natural history and novelty.[34]

However, Smith additionally indicates that the garden/riverbank is not necessarily the place for originality. In the case of "Flora," Smith's notes containing allusions to the verses of other poets occur almost entirely in the poem's first half, which addresses locales according with the garden's familiar and "artificial" scenes. In this context, the garden's artificiality pertains to its associations with cultivation and conventional concerns with aesthetics through overlapping ideas about nature and art; yet the Linnaean taxonomic system that relies on a single character in its classification of species, the reproductive parts of plants, also was thus *termed* "artificial." By contrast, a "natural" system is "discovered" rather than imposed, contains no real exceptions, and is more comprehensive in its methodological observations of any particular species.[35] Significantly, after clarifying her "fabulous" description or "poetical license" regarding the dragonfly, Smith presents botanical knowledge within the garden's familiar and cultivated space as being "well known" and thus briefly suspends the use of scientific notes altogether, claiming that such "notes would be superfluous" when discussing flowers including violets, hyacinths, honeysuckles, and roses. She exemplifies these cultivated garden flowers within a brief "calendar of flora" as representing species that appear successively throughout "the various year" and situates them within the common traditions of classical mythology. Smith thus further emphasizes that this is a space of species that have been thoroughly troped, explored, and documented since the time of ancient writers, so that the location of what is genuinely new and fostering of truth, feeling, and creativity should be sought elsewhere (ll. 99–100).

Poetic Power and Originality in the "Woodland's Wild Uncultured Scene"

Departing from the garden's familiarity, Smith traces Flora to "the deep woodland's wild uncultured scene," a place still presented as being within the knowledge of the botanist (l. 127). Scientific notes return here to expound on trees including the chestnut, larch, and fir, as well as the oak. Resurrecting also reference to botany's military associations through England's imperial power, Smith exclaims that the oak's "giant produce may command the World!," implying naval victory at this time of war against the French, as well as wide-reaching maritime exploration and colonial control through ships constructed with lumber from this tree, the oak, that represents the English nation.[36] Smith thus briefly expands this perspective of place, moving from the woods to the world and back, creating a sense of utility within the vegetable kingdom that was missing from the garden's focus on beauty, wit, and pleasure. The trees and plants of the forest are described in more somber tones as "dark" and "sullen," existing in "rude" "wastes" and "shade" (ll. 131, 137, 138, 136). If Smith tinges the garden/riverbank and mead with hints of problematic possibilities regarding established ideas about marriage, order, and education, she more straightforwardly locates sadness and sensibility in the "umbrageous forest" and "rugged shore," where she places the "careless wanderer," the "Mourner," and perhaps the more serious botanist, as opposed to the botanophile of the garden (ll. 206, 208, 207, 218). The speaker sympathetically hopes that Fancy's visions of botanical species in these environments may bring comfort to those who are "by Fortune, and the World, forgot" and who "mourn / O'er joys and hopes that never will return" (ll. 217, 224). In light of her references to England's imperialism and warfare, it may be worth noting here that Smith's son, Charles Dyer, lost a leg at the Siege of Dunkirk in 1793 and died of yellow fever in Barbados in 1801, embodying for Smith a personal link between England's interventions in the "World" and the resulting sacrifices that must be "mourn[ed]."[37]

Interestingly, it is in this melancholic "woodland" that Smith asserts a dynamic image of her authorial power. After the aesthetic conventions, humor, and intellectual distancing in Smith's portrayal of the garden, the farther that she imaginatively travels from that place and toward the less scientifically established or assured environment of the sea, the more seemingly personal, authentic, and vulnerable her verse becomes. She depicts in the forest a metaphor for the "source" of her poetic creativity (l. 143):

> Ah! yet prolong the dear delicious dream,
> And trace her power along the mountain stream.
> See! from its rude and rocky source, o'erhung
> With female Fern, and glossy Adder's-tongue,
> Slowly it wells, in pure and crystal drops,
> And steals soft-gliding thro' the upland copse;
> Then murmuring on, along the willowy sides,
> The Reed-bird whispers, and the Halcyon hides;
> While among Sallows pale, and birchen bowers,
> Embarks in Fancy's eye the Queen of flowers. (ll. 139–48)

Smith's command to "prolong the dear delicious dream" is, of course, directed inward, continuing this vision through the exertion of "Fancy" or her creative powers, reminding readers that this natural scenery is viewed through the mind's eye, and thus through the power of her memory and imagination (ll. 139, 148). Indeed, when she further exhorts that mental energy to "trace her power along the mountain stream" to "its rude and rocky source, o'erhung / With female Fern," Smith at once traces the vegetable growth nourished by this life-giving source, the mountain stream, and also locates here a feminine emblem of her own creative inspiration in this metaphor of the spring or "source" for the "stream" of her verse (ll. 141–42, 140). She portrays this creative "stream" as "well[ing]" and flowing "slowly," "in pure and crystal drops" that signify the purity and authenticity of her poetic writing, feeling, and knowledge (l. 143). As Daniel Robinson states, Wordsworth's sonnets on the River Duddon reveal the influence of other poets, including Charlotte Smith, especially in his "figurative association of the native or natal river with memory and the use of the river as the symbol for the flow of life" and in emphasizing for the poet the stream's symbolism of "spiritual and creative renewal."[38]

Yet, even as Smith self-reflexively invites herself as well as her audience to "See!" this place of poetic origination that aligns with unmitigated, raw ingenuity, this "rude and rocky source" or creative force that forms the flowing "stream" of her writing gathers momentum through the addition of "murmuring" and "whispers" (ll. 141, 140, 145, 146). In other words, one way to interpret Smith's portrayal of the source and subsequent fruition of her imaginative power is that it begins "pure[ly]" from her own thoughts, her poetic mind, and is then enhanced by the murmuring and whispering voices of other poets and naturalists whose works inform her notes and verse (l. 143). Smith's note to the "Reed-bird," to whom she attributes these "whispers," characterizes this species as "a bird that in a low and sweet note imitates several others, and sings all night" (l. 146). Smith often drew on analogies between birds and bards in communicating the complex relationship between imitation and originality in her verse.[39] As Smith was accused of plagiarism and "imitat[ion]" throughout her poetic career and often acknowledged her borrowings through her notes and prefaces, it is striking to see her apply the term "steals" ("And steals soft-gliding thro' the upland copse"), with its various connotations, while describing her authorial method and arguably alluding to her incorporation of other writers' imaginative and scientific texts as whispers and murmurs that enhance and substantiate her own poetic "songs" (l. 144). Of course, as I discuss in the introduction, intertextuality was pervasive within Romantic-era poetry and does not preclude originality.

At the same time, another (perhaps more rebellious) interpretation of this image of feminine poetic "power" may understand the "murmuring" and "whispers" as representing the form of Smith's own dissent from established poetic and scientific authority. In this case, the challenges she voices in the context of "Flora" start as a small spring that becomes a stream, building into a tributary that creates an ocean of possibilities for new literary, social, environmental, and scientific perspectives. This supports the idea that her questioning of existing forms of knowledge allows her poem to become more potent and original in its continued exploration of place, revealing that certain environments lend themselves more readily to autonomous authorial expression, breaking away from other writers' influence. Thus, as Smith depicts Flora "embark[ing]" in "her light skiff" on this plant-rich stream toward the sea, the poet prepares to leave behind the familiar conventions and authoritative voices within the poem's earlier scope to embark on her own imaginative, scientific path.

Questioning Taxonomies: Placing Ambiguities of the Shoreline

While in the first half of "Flora" the artificial space of the garden/riverbank and mead thrives on conventions and inspires notes that sometimes default to the authority of other writers, Smith becomes increasingly present in the third to last stanza, as the poem's floral train makes its way "Down to the Sea" (l. 165). No longer alluding to the contained, sexualized, feminized space of the garden, Smith journeys to the bold vantage of a cliff's summit—a location that clashes with female propriety while her notes take on a more personal air.[40] Depicting the summit's surrounding scene, she draws attention to "the light Tamarisk" that dwells "half way up the clift" and "hangs o'er the ever toiling Surge below" (ll. 171–73). Implying that she has scaled these dangerous climbs to see the plant with her own eyes, Smith catches the pun of the plant's name; she has taken the "risk" of exploring these cliffs for herself. And, lest there be any doubt that the poet speaks from personal experience, the note to "Tamarisk" provides additional proof of Smith's presence: "This elegant plant is not very uncommon on cliffs in the West of England, and was in 1800 to be found on an high rock to the Eastward of the town of Hastings, in Sussex." Specificity of year and location validate and personalize the poet's observations, and Smith's earlier sonnets provide precedence not only for situating the poet herself "in Sussex," but also for envisioning her as a frequenter of cliffs.[41]

As numerous scholars have discussed with regard to the prospect poem, "the summit" typically denotes the vantage of a male poet.[42] Smith, aware of her break with gender conventions, mitigates the potential subversiveness of the situation by depicting a fisherman at the cliff's edge in her stead. Still, the implication remains that Smith has occupied this masculine space, and the difference between *her* relationship with the environment and that of the fisherman stands out in relief as she portrays where

> the summit bare
> Is tufted by the Statice; and there,
> Crush'd by the fisher, as he stands to mark
> Some distant signal, or approaching bark,
> The Saltwort's starry stalks are thickly sown,
> Like humble worth, unheeded and unknown! (ll. 173–78)

In the note to "Statice," Smith explains that the plant "is frequently used for borders of flower beds. It covers some of the most sterile cliffs" (l. 174n). Just as "Statice" exists on the "borders" between land and sea, between the familiar and unfamiliar, as well as between the cultivated and wild realms of flora, Smith presents herself as occupying the borderland between masculine and feminine expectations of the poet, a duality or placement that also arguably exists for her writings that meld technical science and poetry.

Smith indicates the liminality of her own authorial position, in part, by identifying more strongly with the natural objects of "Flora" as the poem delves more deeply into the realm of ambiguity. In contrast to her minute observations of the natural world in her verse and notes, she depicts the "fisher," who "mark[s]" a "distant signal" or ship, but fails to notice the plant, Saltwort.[43] Smith's note for Saltwort explains that the plant "is used in the manufacture of glass. The best is brought from the Mediterranean, and forms a considerable article of commerce. It is very frequent on the cliffs of the Sussex coast" (l. 177n). Smith thus exhibits not only her knowledge of the fisher's own terrain, and the possible origination and value of shipments for which he peers so intently, but also her moral superiority over the fisher through her greater sensitivity to life, expressed here as superior care and observational skills. It is tempting to suggest that this fisher represents Smith's parody of contemporary male poets who stand at lofty heights and channel their mental powers to squint into an abstract distance. As Lisa Ottum writes of this author's fictional landscapes, Smith paradoxically "associates panoramic scenery and the masculinized posture that accompanies it with shortsightedness."[44] Too absorbed in some faraway object to notice the intricate worlds of which he forms a part, the fisher "crush[es]" the Saltwort underfoot.[45] Smith's depiction of Saltwort's "starry stalks" and of its role in the production of glass amplifies the plant's delicate fragility. Saltwort's association with glass additionally conjures up various tools of observation such as the microscope and magnifying glass so that the plant arguably participates in its own discovery and inspection, a self-reflexivity that extends to the poet as well. The Saltwort is another native of Sussex, and one can hardly fail to hear an analogy when Smith likens the plant to "humble worth, unheeded and unknown!," projecting her own experience of feeling, like this plant, "crush'd" and unnoticed while in a vulnerable financial, physical, and emotional state.[46] Still, in this moment, as the fisher "mark[s]" "some distant signal," Smith wishes the reader to mark *her* signal, warning of the poem's immediate destination. Not only does proceeding past the stanza's edge mean stepping off the cliff's edge to plunge into the sea but, in taking that step, we also enter the realm of the seemingly unorderable, unidentifiable, "unknown!"

In the poem's penultimate stanza, Smith's verse descends into the "depths" of the sea and brings to light the ineffectiveness of classification systems that cannot tolerate ambiguity. Here, she presents the sea as an environment teeming with various liminalities between kingdoms, classes, orders, and species that set the taxonomist—for whom everything must fit into its single, designated category—on edge. In this way, Smith depicts knowledge and certainty in crisis, and the sea as a place for the placeless. Gazing on the gradually receding shoreline of the known, the poet-naturalist provokingly spotlights oceanic objects that disrupt botanical order. Smith sets the stage for this taxonomic disorientation with "corals," which she uncharacteristically denies a note, perhaps because she takes for granted that the dispute over whether to classify corals within the plant or animal kingdom is well known (l. 179). Barbara Stafford explains that corals and ceratophytes "decentered both kingdoms by calling any hard and fast boundaries between them into question. . . . These living ambiguities obeyed William James's 'law of dissociation,' whereby what was first associated with one thing and then with another tended to become dissociated from either and grew into a separate object of contemplation."[47] Corals thus become the first of this stanza's many border-lives, elucidating organisms that classification struggles to encompass.

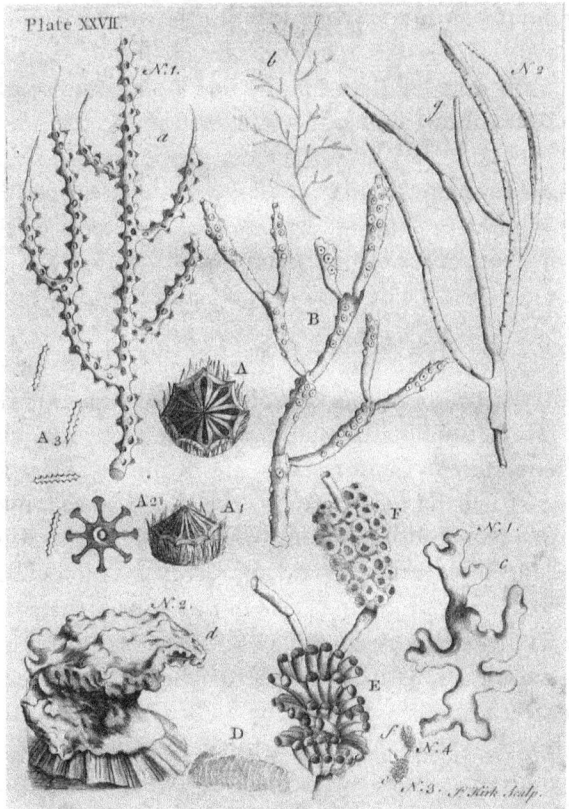

FIGURE 9. Plate 27 of John Ellis's *Essay Towards a Natural History of the Corallines* (London, 1755). (Image courtesy of the Linda Hall Library of Science, Engineering, and Technology)

Following her inclusion of corals, Smith's verse provides several additional examples of species that classification strains to contain. At the same time, her undermining of the scientific taxonomies on which she built authority in the poem's first half risks forfeiture of her own claims to knowledge as well. Her note for "Algae" deploys the pronoun "I" (only Smith's second use of this pronoun in "Flora").[48] Yet, rather than bolstering the poet's authority, the self-reference suggests incertitude due to shortcomings in the ordering system on which she relies. In what sounds like a parody of the taxonomic technique of "lumping," which privileges organisms' commonalities over differences, Smith's note hazards that "*Algae, Fuci* and *Conferva*, include, I believe, all sea plants" (l. 181n). The disclaiming phrase "I believe" fragments Smith's tone of confidence in her scientific knowledge. Becoming increasingly obvious in her exposé of the unknown, Smith next exhibits the "Polyp," which, like coral, threatens with liminal status between kingdoms, balancing on the border of "half flower, half fish" (l. 182).[49] Not confined to the note, this insertion of ambiguity directly into the poem's verse increasingly registers fractures in the taxonomic order. Fracturing further, in the note for "Coralline," Smith questions both her source of information and her ability to decipher its meaning: "Coralline is, if I do not misunderstand the only book I have to consult, a shelly substance, the work of sea insects, adhering to stones and to sea weeds" (l. 184n). So phrased, Smith shifts blame for uncertainty away from herself and onto this lack of textual information that thwarts her efforts at exactitude and thus emphasizes the need for firsthand observations. In her preface to *Conversations*, Smith admits, "I fear I have made some mistakes, particularly in regard to the nature of zoophytes; but the accounts of this branch of natural history in the few books I have are so confused and incompleat that I could not rectify the errors I suspected."[50] Smith likely references Coralline algae in this part of the poem, and she frustratedly attributes the possibility of scientific "errors" regarding plant-like animals, such as corals, in her work, to naturalists' failures to communicate or discover necessary information. In her verse, even language itself now strikes Smith as inept, prompting further endnoted apologies for her use of "Panier'd": "Panier'd is not perhaps a word correctly English, but it must here be forgiven me" (l. 191n). It must be forgiven because correct definitions and clarity within systems of language and identification are quickly eroding in the poem's context.

Abandoning Taxonomy:
Describing the Sea's "Unknown"

In her penultimate stanza, Smith makes one last effort to salvage systems of knowledge before taxonomy collapses entirely within the poem's structure. In her note for "Pinna," or "the silk-worm of the sea," she asserts confidence, but gives up authority by predicating her knowledge on a note (and thus a scientific, not poetic, positing) of Erasmus Darwin. This, Smith's final note of the stanza, proves pivotal. After symbolically abandoning her own attempts to make sense of taxonomic uncertainties, she announces the implementation of an alternative method, declaring, "*The subsequent lines attempt a description of sea plants, without any correct classification*" (l. 192n; emphasis mine). While Smith's phrasing obscures whether these sea plants have no "correct classification," or that she simply lacks the desire and/or means to classify them, her preceding taxonomic frustrations support conjoining these two interpretations. The poet cannot classify the remaining sea plants because she has shown taxonomies to be unreliable within this biogeographical space of broken boundaries. In the face of the "unknown!," naturalists' systems of knowledge have failed her. Thus, Smith abandons taxonomy and retains only natural history's emphasis on observation. She revels in revealing species that apparently have no place within naturalists' extant ordering systems, the unidentifiable, the unnameable, that which can only be communicated through "description" and thereby evades being subject to (re)interpellation through classification's impulse to claim and (re)name.

In practice, Smith's verse models potential for simultaneously preserving the unfamiliar and familiar aspects of species through poetic description. While doing so, she additionally feminizes and mythifies the ocean's alien realm, populating it with "Sea-maids" who respond to the female Fancy's call:

> Each her trophy brings
> Of plants, from rocks and caverns sub-marine,
> With leathery branch, and bladder'd buds between;
> There its dark folds the pucker'd Laver spread
> With trees in miniature of various red;
> There flag-shaped Olive leaves depending hung,
> And fairy fans from glossy pebbles sprung. (ll. 192–98)

These lines, detailing "sea plants" Smith encountered on excursions along the Sussex coast, express her mind's free associations based on sensorial experience. While proclaiming the dictation of Fancy, she communicates through her observations all that can be known about these natural objects. Although "Laver" is a common name for edible seaweed, Smith's location of "Olive leaves" in the sea may seem initially puzzling, especially as she discusses in *Conversations* the cultivation of olives and includes an "Ode to the Olive Tree," displaying her knowledge of this Mediterranean plant.[51] However, she likely intends the "olive sea thong," a brown- or olive-colored seaweed found in the Atlantic Ocean that she references in her poem "An Evening Walk by the Sea Side," also included in *Conversations*.[52] Vegetable analogies relate the unknown to the known, conjuring up familiar plant imagery ("branch," "buds," "trees," "leaves"), while the poet relies upon carefully crafted modifiers to simultaneously illuminate the scene's foreignness ("leathery," "bladder'd," "pucker'd," "various red"). Through her method of description, Smith displays how these natural objects in all their variance can be appreciated, and not merely defined or delimited. Celebrating nature in the absence of names and orders, the poet-naturalist gazes in sympathetic identification with the unidentified, whose very existence retrospectively destabilizes Smith's early confidence in the poem's classifications. In this way, she not only interrogates naturalists' assertions but also seeks more accurate information and poetic originality through the depiction of that which is "unknown."

In this stanza, Smith thus forfeits her (taxonomic) knowledge, but regains her authority, which she upholds with her personal observations; even common names no longer suffice as Smith forces herself and the reader to look at the world with new eyes. At the beginning of the poem, Fancy aligned with taxonomy and convention, but, in discrediting aspects of contemporary classification, imagination now conjoins with descriptions of nature—not racing to reduce, trope, and systematize, but appreciating a boundlessness to the observation, to what is produced in the imagination, and to the organism itself. Employing descriptive methods, Smith nevertheless recognizes that even this component of empiricism is often inadequate, due to limitations of language, experience, and the "human eye," as she writes in a note about the landscape in "Beachy Head": "Description falls so infinitely short of the reality, that only here and there, distinct features can be given" (238). In the notes to another poem, she remarks on her propensity for asking "questions which I have generally been stared at for making."[53] In light of her fearless questioning of taxonomies, it is notable that it was not until many years after the publication of Smith's "Flora" that the sea's ambiguities received extensive scientific attention. In the 1840s and 1850s, in particular, Philip Henry Gosse, William Henry Harvey, and David Landesborough each introduced to popular audiences the life of the seashore, helping to create marine biology.[54]

In her sea exploration, Smith's focus on those biological forms that challenged the classificatory systems of her day provocatively gestures toward mysteries within the natural world. In this vein, she ends the poem by reemphasizing observation and "description" within contexts of sensibility, transforming the figure of the "Mourner" into a melancholy naturalist who looks steadily at botanical objects as well as at experiences of sadness to pursue some of life's uncertainties. Flowers here "soothe the wearied

Pilgrim's eyes" and "afford an antepast of Paradise," so that the mourner's vision or "eyes" seem "wearied" both from crying and from closely studying and examining plant specimens for emotional comfort. This interpretation is also supported by similar imagery in Smith's sonnet "To the Goddess of Botany," which in many ways reads as "Flora" in miniature, where she writes of botanical studies as "soothing" for "my tired, and tear-swoln eyes" and notes, "I cannot now turn to any other pursuit that for a moment soothes my wounded mind," for "it has been my misfortune to have endured real calamities that have disqualified me for finding any enjoyment in the pleasures and pursuits which occupy the generality of the world" (68–69). In that sonnet, botany serves as a form of escapism and the speaker's desire for "silen[ce]," and "rest" sounds like a death-wish, bringing potential reference to suicide by drowning through the pursuit of plants that participate in her sadness with "trembling sighs" and finally lead her "beneath the Ocean waves" (68–69). Although "Flora" does not suggest the Mourner's death, there is clear consideration of the afterlife. In Smith's work, attention to "unknown" botanical species enables the melancholy naturalist to balance, on the one hand, spiritual mysteries about why suffering occurs in the world and, on the other, efforts to solve nature's mysteries with greater empathy by understanding species more fully than, say, contemporary artificial systems of classification allowed. The Mourner's botanical studies thus serve as an "antepast of Paradise" through implied reference both to the afterlife's recompense for earthly suffering, and the Christian belief that, in Heaven, answers to life's material and emotional uncertainties will be revealed, will become known.[55]

Smith's engagement with natural history throughout her poetic career suggests that her questioning of taxonomic systems is not a rejection of science as a means to knowledge, but rather a call for more careful attention to various natural environments and the species that inhabit them, with an approach that is more inductive than deductive. After all, in her *Natural History of Birds* (1807), she declares, "the philosopher and poet should both be naturalists."[56] Nevertheless, even as she argues for greater understanding of the natural world, she highlights ways in which the unknown possesses its own beauty and power, with sociobiological implications. In this way, depicting her writing career in the poem "To My Lyre," Smith delineates women who encode the social construction of femininity, but then quickly demands a broader taxonomy than that which creates a mere gender binary, "For," she declares, "*I was of a different species*" (ll. 19–22, 24; emphasis mine). As the self-reflexive contexts of "Flora" suggest, similar to the less conventional species Smith selects, women writers who defy the gender expectations of analogously becoming the cultivated flowers of the garden sustain an unknowability, if also a kind of marginality. These women create space for subversive modes of insight and authority as well as multiplicity in regard to ideas or embodiments of identity. By pointing out organisms that defy classification, Smith indicates the constructedness of taxonomic orders, and, in destabilizing categories imposed on botanical specimens, she arguably seeks to legitimate a freer space for women (especially women writers) as well. Her life and work may be richly described, but not easily placed or confined within a particular set of social values or expectations. Juxtaposing her own kind of border-life with other border-lives in nature, Smith exposes, in Jacqueline Labbe's words, the "constructedness of the margin" and productively disputes the constrictive categories and knowledge-base on which ordering systems depend.[57] She thus opens possibilities for mysteries of place, with all its connotations, to be explored through direct knowledge, observations, and descriptions that provide empathy and respect for difference as a means both to poetic originality and new modes of scientific, political, and literary power within sensibility. However, it is a very different conception of "power" from that which is later conceived by Wordsworth and De Quincey (and that I discuss in the conclusion).

While, in this chapter, Smith employs botany and sensibility to critique perceived shortcomings in the works of contemporary male poets and naturalists as well as in scientific taxonomies, my final chapter explores how Anna Seward seeks to vindicate sensibility itself from negative portrayals within certain medical and botanical texts. Eighteenth-century medical treatises, including Darwin's, often viewed excessive sensibility as engendering forms of insanity. Contrasting this pathologization of sensibility, Seward instead asserts that it is not the possession, but the absence, of sensibility that should be associated with madness. She thereby suggests Darwin's own madness, delineating his and his botanical poetry's lack of sensibility. Moreover, Seward similarly aligns both William Cowper and William Wordsworth, as well as some of their botanically related verses, with madness, due to their corruptions, in her view, of sensibility through mysticism and egotism. And, as I show, her engagements with this poetic and scientific discourse of madness and sensibility also interestingly connect with John Clare's distrust of Linnaean botanical taxonomy.

6

Botany and Madness

Anna Seward, Sensibility, and the Floral Insanities of Darwin, Cowper, Wordsworth, and Clare

ENTHUSIASM FOR botany during the Romantic period reached such a height that, according to some scholars, it constituted a kind of madness, or *botanomania*, a term reminiscent of the tulipomania that swept the Netherlands in the seventeenth century. Indeed, in this later era, there arguably existed in concepts of literature a triangulated relationship between botany, madness, and originality. Notions of madness and genius or originality had been connected in literary and philosophical discourse since classical times in the works of Plato, Aristotle, and Seneca, among others.[1] Botany and madness had also been linked in various ways for centuries through literature, medicine, and analogy.[2] The discoveries of "new" botanical systems and species held the promise of originality in both science and literature through the communication of new information and ways of thinking about the natural world. Nevertheless, at the same time that Linnaean taxonomy promised to impose order and stability, it also vied with various other attempts at plant classification so that when Robert Thornton published his popular variation of the Linnaean system in 1799 he counted fifty-two different systems of botany, causing one botanist to complain of "system-madness."[3] As James Whitehead notes, "In the nineteenth century, 'madness' was (and still is) a highly dynamic and mobile term which covered a wide range of bodily, behavioral, and environmental conditions," while "insane" or "insanity" can be used when "a more explicitly pathological sense is intended."[4] Mid-eighteenth-century literary critics also sometimes associated genius with plants and sanity, for, as Roy Porter explains, "while prizing originality, William Sharpe's *Dissertation upon Genius* (1755) and Edward Young's *Conjectures on Original Composition* (1759) read creativity as the outpourings of the healthy psyche, analogous to the growth and flowering of plants"; such critical assertions established connections that would be repeated throughout the Romantic era in, for example, Charles Lamb's essay "The Sanity of True Genius" and Coleridge's organic theory analogizing plants and poetry.[5]

As I show in this chapter, Anna Seward, the literary critic and poet of sensibility and often sociability, employs concepts of sensibility to interrelate and interrogate botany, poetry, and madness; and she does so in ways that transform our understandings of both her works and those of other poets including Erasmus Darwin, William Cowper, William Wordsworth, and John Clare. As I argue in this book's introduction, Wordsworth and other contemporaries accused Darwin of an incapacity for sensibility that then became imposed on scientific poets more generally, creating an inaccurate conception of the period's scientific literature. In fact, other writers, including Seward, who incorporated both sensibility and natural history into their literature also sometimes faulted Darwin as unfeeling. In her *Memoirs of the Life of Dr. Darwin* (1804), Seward ostensibly confines her criticisms of Darwin's medical treatise, *Zoonomia* (1794–96), to his section, "Of Instinct"; however, as I will demonstrate, she also subtly appropriates from this text Darwin's scientific theories of insanity and employs (or alters) them in her analysis of his poem *The Botanic Garden*, and elsewhere in the *Memoirs*, to suggest Darwin's own madness, particularly due to his lack of feeling, "sensation," or sensibility. Thus, Seward defends sensibility from its negative portrayals in eighteenth-century medical treatises, where madness is often associated with sensibility "in excess." Whereas Darwin draws on medical writings and "nerve theory" to assert such connections between sensibility and insanity, Seward's conception of sensibility challenges and reframes what can or should be classified as madness. Moreover, I argue that Seward continues to rework and apply Darwin's hypotheses about insanity in her published letters regarding not only Cowper, whose struggles with mental health were well known, but also Wordsworth, critiquing both men's "egotism" and "mysticism" within botanical contexts that also interestingly connect with Clare. Through strategic accusations of madness within her literary criticism, Seward makes the case that both Darwin and Wordsworth fail to appropriately practice sensibility in their floral poems, and, in doing so, she instead champions poetry that can be associated with sociable, connective forms of sensibility. Within its conception as an inherently solitary and isolating experience, madness signifies in Seward's view the danger of developing, Romantic-era concerns with subjectivity and individualism in their inaccessible and alienating extremes. Placing herself in the powerful position of literary critic in a period when the line between science and literature was often blurred, Seward assumes the ability to contribute to and assess both fields in relation to their respective associations with in/sanity, while also seeking to vindicate and delineate sensibility to stress its continued importance.

Seward, Literary Criticism, and Botany

Much of Seward's literary criticism is contained in her *Memoirs* on Darwin and in her published letters, and she discusses botany on numerous occasions in both of these works as well.[6] I have argued elsewhere that, in Seward's methods of literary criticism, she often functions as a "literary naturalist."[7] In this way, she classifies poets into hierarchical literary taxonomies that display their relative success within a particular verse form and their relations to other writers, and she details the particular writing styles of individual poets as minutely as a naturalist might describe a natural object. In addition to discussing botany in these texts, she also sometimes incorporates references to this and other fields of natural history in her poetry as, for example, in her celebrated *Elegy on Captain Cook* (1780), which specifies in prose footnotes some of the exotic biological species recorded during James Cook's voyages. As a literary critic, Seward placed herself in a position of judgment, and thus authority, in relation to other writers. She also gained numerous correspondents, many of whom were fellow poets, and enjoyed mentoring younger writers as a means of shaping the future development of poetry; of course, her published critiques of other writers had this goal of influencing posterity as well.

Undoubtedly, Seward's interest in botany owed much to the company that she kept; in addition to Darwin, she spent time with John Saville, a vicar-chorale of Lichfield Cathedral who began giving her music lessons in the 1760s and became her close friend, and "probably [her] inamorato," until his death in 1803.[8] She describes one instance in which she, Darwin, and Saville were walking to Saville's garden while discussing "some rare and beautiful plants" and, on Seward admitting that she had never seen the *Calmia*, Darwin playfully remarked that "it is a flower of such exquisite beauty, that would make you waste the summer's day in examining it," and its color is "precisely that of a seraph's plume," causing them all to laugh "at the *accuracy* of the description."⁹ Alluding to Saville's "botanic enthusiasm," Seward admires that "the botanists all love each other the better for the knowledge and vegetable treasures that each possess" and contrasts this with the "envy" often present among poets, wondering, "Why do not the bards thus also?" (*Letters*, 1:170–71). With obvious pride in Saville's botanical achievements, Seward wrote in a letter to Helen Maria Williams in 1788 that he "is engrossed by attendance upon at least two thousand rare plants and flowers, so that his friends lose many hours every week of his company;—hours which they do not like to spare. But his fame as a botanic florist flies far. On the side of Johnson's favorite gigantic willow . . . lies his little garden. It is become one of the Lichfield lions which strangers go to see" (2:180–81). She reports that Saville declined to join the Lichfield Botanic Society, which included Darwin and two other members, to avoid the burdens of their efforts to translate Linnaeus's works (6:175–76). Seward confesses, "I was always fond of flowers; but, perfectly satisfied with the great variety of tints and odors which those of a common garden supply," and thus "secretly deplored, as the misfortune they really were to him, [Saville's] botanic knowledge and thirst for so immense a possession of rare plants and flowers;—since their culture was difficult, troublesome, and expensive, engrossing a great deal of time which, his admirable talents considered, might have been better employed" (6:177).

Regarding the versification of natural history, Seward writes that she agrees with literary critics including John Aikin and Thomas Percival in encouraging poets to achieve originality through engagements with these scientific fields, and in their "general censure of the poetic violations of natural history" (*Letters*, 1:18–20). While she holds James Thomson and other poets accountable for their botanical inaccuracies, she still specifies that the pursuit of accuracy should not impede poetic expression (1:18–20). Referencing Darwin's *Botanic Garden* and Aikin's *Essay on the Application of Natural History to Poetry*, she also defends the inclusion of informative prose scientific notes where they are useful in poetry (1:73–74). In a statement aligning with the exhortations in Aikin's *Essay*, Seward writes, "It has always been my endeavor to paint from nature, rather than to copy from books, in my poetic landscapes" (1:219). She also had an acquaintance with Sir Joseph Banks and patriotically politicized styles of gardening, suggesting that "the gardens of England are as much more natural, free, and beautiful, than those of any other country, as its constitution diffuses more genuine liberty" (2:31; 4: 10).

Nevertheless, Seward developed a complicated relationship with Darwin and his botanical poetry. On numerous occasions in her published letters she praises his *Botanic Garden* and defends it from the critiques of her correspondents.[10] She asserts the propriety of young women reading it, stating that "Dr. Darwin's subject is a real operation of nature, and of nature's God, in the unaltering laws of his vegetable world" (*Letters*, 6:142–43). Seward writes that Aikin "observed, in his dissertation on the subject, that the union of natural history and of modern philosophic science with poetry, was the desideratum in the fanes of the muses" and "Darwin's great and splendid poem supplies this desideratum" (6:157). She also assures Darwin that Saville and other friends "continue to explore with me the poetic graces of the *Botanic Garden*, with delight 'which grows by what it feeds upon'" (2:278). However, later in 1789, she explains to another friend that "some lines of mine, about fifty in number, had the honor of suggesting to Dr. Darwin the first idea of the" *Botanic Garden*, and these lines were published in the *Gentleman's Magazine* in May 1783 in her name, prompting her to condemn his plan to appropriate these lines for the opening of his *Economy of Vegetation*: "Surely he judges wrong: so great a work ought not to contain lines, especially in the exordium, which are known to have been written by another" (2:311–13). After Darwin published the first part of his poem, including her lines, Seward periodically continued to repeat this accusation in her correspondence, and also in her *Memoirs* of Darwin, stating of the lines in his exordium, "Four-fifths of them are mine verbatim, and mine the whole order of the scenery, so that a charge of plagiarism must rest somewhere" (3:155–56). In one letter she writes that Darwin "became a sort of poetic preceptor to me in my early youth. If I have critical knowledge in my favorite science [poetry], I hold myself chiefly indebted for it to him" but, prior to *The Botanic Garden*, Darwin never "venture[d] to appear before the world as a bard" for fear that it may harm his medical practice (2:312). Seward complains, "Since he commenced poet professed, Darwin is become notoriously guilty of the narrow-souled jealousy. Till then he was a warm admirer and generous encomiast of poetic effluence, in whatever form it might appear" (3:187).[11] She faults Darwin's *Botanic Garden* for its "irreligion, and encomiums on the terrible and tyrannic democracy of France" (5:115). And Seward took pleasure in George Canning's *Loves of the Triangles* (1798), published in the *Anti-Jacobin*, calling it an "exquisite satire on the plan and on the absurdities of Darwin's supremely ingenious, but very affected [*Botanic Garden*]" (5:113). Still, she writes that this satire does not lessen her appreciation of Darwin's poem "because I never considered that poetry as faultless, or its style as the best model for rising genius to adopt" (5:136). As I discuss below, Seward's critical judgments of Darwin's poetry must be understood within the context of contemporary conceptions of sensibility.

Darwin's Madness: "Nerve Theory" and the Lack of "Sensation"

Scholars have frequently been puzzled by Seward's treatment of Darwin and his writings in her *Memoirs*. Desmond King-Hele, for example, acknowledges that the *Memoirs* is "often quite waspish," but also "discerning and magnanimous."[12] After reading Seward's text, rival poet Charlotte Smith compared her with a "jackal at prey," feasting on Darwin's personal reputation and literary remains.[13] In her published letters and in the *Memoirs* itself, Seward defends her portrayal of Darwin as a balanced and realistic account of this well-known poet, physician, and naturalist (*Letters*, 6:73–74, 136–37). However, I argue that, in her *Memoirs*, Seward also subtly applies Darwin's theories from *Zoonomia* in her critiques of his botanical poetry and personal conduct to display his lack of sensibility and imply that he could, by his own standards, be categorized as insane. By suggesting Darwin's madness, she additionally defends her own conception of poetry in conjunction with her literary criticism and standing as a poet of sensibility.

In the *Memoirs*, Seward repeatedly accuses Darwin's *Botanic Garden* of inadequate "sensation." Sharing responsibility for this "just and delicate criticism," Seward identifies the censure as originating with her friend and correspondent, Robert Fellowes, who states, "Dr. Darwin's poetry wants sensation; that sort of excellence which, while it enables us to see distinctly the objects described, makes us feel them acting on the nerves" (170–71). Demonstrating this dearth of sensation in Darwin's poetry, Seward first compares verses from *The Botanic Garden* with a sonnet by Charles Lloyd, each of which describe "a wintery evening, in late autumn" (171). She explains that Darwin's poem, "though more dignified, does *not* thrill our nerves, and the second *does*. We admire in [Darwin] the power and grace of the poet; in [Lloyd] we forget the poet and his art, and only yearn to see images reflected in his mirror, which we have annually, and many times shuddered to survey in real life" (172). Seward writes that *sensation* "seems a new term in criticism" and differentiates it from the strong passion of *pathos*, which acts on the heart, delineating that sensation primarily affects the nerves, representing a "gentler, subtler, and more evanescent influence, which almost imperceptibly touches the passions without agitating them" (176–77).[14] By noting in Lloyd's work the capacity for sensation, Seward imbues it with the ability to transcend mere observation, activating feeling through readers' identification with the described experience. Significantly, Seward and Darwin personally knew Lloyd, who was also friends with Coleridge, Lamb, and Southey, as Lloyd spent time under Darwin's care in his sanitorium in Lichfield.[15] Although Seward never mentions Lloyd's struggles with mental health, her comparison emphasizes his emotional adeptness in contrast with Darwin's dearth of sensation.

Indeed, Seward aligns Lloyd with Shakespeare, whom she felt to be the epitome of British poetry and through whose works, again, "we feel sensation," and she ascribes this quality to Milton and Gray as well (173, 175, 179). She likely felt prompted to these comparisons not least because at the height of Darwin's fame he was "thought by many to be the equal of Milton or Shakespeare," so that John Keats later scoffed at admirers of poetry who "jumble together Shakespeare and Darwin."[16] According to Seward, "Probably the reason why Dr. Darwin's poetry, while it delights the imagination, leaves the nerves at rest, may be, that he seldom" incorporates the "*moral epithet*, meaning that quality of the thing mentioned, which pertains more to the mind, or heart, than to the eye, and which, instead of picture, excites sensation" (173–74). She thus denigrates his emphasis on "the eye" and visual objects in his poetic theory and methods as described in the prose interludes of *The Loves of the Plants*. So, unlike Darwin's "poetic" "picture" of a glowworm, Shakespeare's verses about this insect provide "no distinct picture" and "is not descriptive" because its "moral epithet, *ineffectual*, does better than paint its object" and "excites a sort of tender pity for the little insect, shining without either warmth or useful light, in the dark and lonely hours" (174–75). Culminating her critique of Darwin's poetry, Seward writes that his "excellence consists in delighting the eye, the taste, and the fancy, by the strength, distinctness, elegance, and perfect originality of his pictures; and in delighting the ear by the rich cadence of his numbers; *but the passions are generally asleep, and seldom are the nerves thrilled* by his imagery" (177; emphasis mine). In contrast with Britain's greatest poets, Darwin "would not have succeeded so transcendently on themes, which demanded either pathos, or that sort of tender and delicate feeling in the poet, which excites in the reader sympathetic sensation; or yet in the sacred morality of ethic poetry" (179).

Seward thus signals to "the reader" that the capacity for "transcenden[ce]," "pathos," "tender and delicate feeling," "sympathetic sensation," and "morality" are all necessary for superior poetry, and are beyond Darwin's poetic reach. Relegating Darwin's poetry to its appeal only for "the eye," while the preeminent English poets access the nerves, "mind, or heart," Seward, as a literary naturalist, precludes him from being categorized with those literary greats and further defines what she means by "sensation."

Seward strategically emphasizes Darwin's poetic shortage of "sensation," or his inability to act on the nerves, as a means of responding to eighteenth-century medical treatises and conceptions of "nerve theory." As I have shown elsewhere, in the *Memoirs,* her section designated for discussion of Darwin's *Zoonomia,* focuses solely on what she views as his misrepresentation of "instinct," which Seward interprets as also having significance for literary authorship and the cohesiveness of poetic identity.[17] Fascinatingly, without specifically acknowledging that she is doing so, Seward continues to draw on Darwin's *Zoonomia* throughout her analysis of his botanical poetry and personal life. Anyone who had read *Zoonomia* would recognize "sensation" as one of his key terms. In that medical text, Darwin designates the human body's "living principle, or spirit of animation," as the "sensorium," which may also refer to the medullary part of the brain, spinal marrow, nerves, organs of sense, and of the muscles (1:10). According to Darwin, the sensorium has four different faculties or "modes of action": irritation, sensation, volition, and association (1:32). He writes that the "contraction of animal fibres" producing pleasure or pain comprise sensation, and "sensation produces desire or aversion," which constitutes volition through "efforts of will" (1:30, 34). Also, while sensation refers to this faculty in its active state, its inactive state is termed *sensibility* (1:32). Seward's particular attention to "nerves" thus has medical basis as eighteenth-century physicians often discussed sensibility in terms of nerve theory and as "nervous disorders," describing nerves as connecting the mind and body so that a disorder of the mind often manifested itself with bodily symptoms; it was thought that "delicate nerves" could signal the capacity for superior intellect and feeling, as well as morality, but such disorders could also pass into melancholy or lunacy and were considered "very common among the Literati."[18] At the same time, well-regulated "sensibility was thought to arise from and to signal a proper and ordered working of the nerves, . . . a healthy and organic body derived from healthy organs of feeling."[19] Moreover, as one eighteenth-century physician asserted, "all sympathy, all consensus presupposes sentiment and consequently can exist only by the mediation of the nerves, which are the only instruments by which sensation operates."[20] Such statements enabled the culture of sensibility to claim scientific basis for promoting "an image of society held together by bonds of feeling."[21] Since Seward became famous in the last decades of the eighteenth century as a poet of sensibility, Darwin's employment of the concepts of sensation and sensibility in his medical treatise clearly caught her attention as providing additional overlaps with the realms of literature and literary criticism.[22] Moreover, and importantly, Seward took offense at Darwin's participation in the medical appropriation of these terms and their applications in contexts of determining insanity.

Indeed, in *Zoonomia*, Darwin declares that his patients' constitutions that prove "most liable to insanity, are such as have excess of sensibility" (1:354). He also classifies various "species" of insanity within a taxonomy for "diseases of volition" (1:356). Among these insanities, he includes certain kinds of poetry, designated as "sentimental love," wherein "the object of love is beauty" (1:363–64). If a "mad poet" could be said to possess, in Darwin's assessment, an excess of sensibility, then, as Seward proceeds to show, a *lack* of sensation or sensibility is equally a sign of insanity, and it is this dearth of feeling that she pinpoints as the cause for both his personal and poetic failings. In fact, Darwin himself states that "in the minds of mad people those volitions alone exist which are *unmixed with sensation*," suggesting that excess in either direction of possessing or lacking sensibility constitutes madness (1:433; emphasis mine). Seward thus sets up a paradigm in which Darwin, who was a member of the Lunar Society of Birmingham, where participants engaged in inquiries regarding the natural sciences and industrial advancement and called themselves "Lunaticks," could be aligned with the stereotype of the "mad scientist" figure due to his want of human feeling, and emphasizes his poetic devotion to "the eye" rather than to "the heart" as "a radical defect in his poetic system" (*Memoirs*, 307).[23]

In this vein, the closing pages of Seward's *Memoirs* bring some of her most shocking critiques of Darwin's personal character and reinforce her depiction of him as unfeeling. Just as she suggested that Darwin's lack of sensation constitutes a "defect" in his verse, she likewise implies that it creates a parallel defect in his personality, indicative of his possible insanity. Perhaps the most startling example of this is her account of Darwin's reaction to the suicide of one of his sons, who was also named Erasmus. According to Seward, Darwin's son "became the victim of secret and utter despair," for "any more than ordinary recurrence of professional business perplexed and oppressed him" (404). In her account, on a "cold and stormy" evening in December 1799, the "river Derwent, which ran at the bottom of his garden, was partially frozen" and Erasmus junior "was on the couch complaining of the head-ach" when his business partner went out on an errand (405). Upon the partner's return, he found Erasmus's hat and neckcloth in the garden "walk which leads to the river" so that an

> alarm was immediately given, and boats were sent out. Dr. Darwin had been summoned. He staid a long time on the brink of the water, apparently calm and collected, but doubtless suffering the most torturing anxiety. The body could not be found till the next day. When the Doctor received information that it was found, he exclaimed in a low voice, "Poor insane coward!" and it is said never afterwards mentioned the subject. (406)

Seward describes Darwin's son as benevolent, "beloved, respected, and mourned by all who knew him" (406–7). Although Erasmus shared his father's name, Seward makes it clear that he did not share his personality, relating that, when younger, this son had

> expressed a wish to go into the Church rather than the Law. That preference was repulsed by paternal sarcasms upon its indolence and imputed effeminacy. From infancy to his last day, Mr. Darwin had shrunk, *with pained sensibility*, from his father's irony. Probably from the less active, less scientific disposition of Erasmus, in comparison with that of his brothers, Charles and Robert, Dr. Darwin had always appeared colder towards him than to his other children. Doubtless it was that inferior degree of attachment which made the lesson of stoicism somewhat more practicable on this trying, this dire occasion. It excited, however, universal surprise to see him walking along the streets of Derby the day after the funeral of his son, with a serene countenance and his usual cheerfulness of address. This self-command enabled him to take immediate possession of the premises bequeathed to him; to lay plans for their improvement; to take pleasure in describing those plans to his acquaintance, and to determine to make it his future residence; and all this without seeming to recollect to how sad an event he owed their possession! (407–9; emphasis mine)

Interestingly, Seward's reference to "the lesson of stoicism" recalls Darwin's *Zoonomia*, in which he taxonomized as insanity the excessive experience of "grief," or "a perpetual voluntary contemplation of all the circumstances of some great loss, as of a favorite child" (2:370). Countering this assertion, Seward instead highlights Darwin's *lack* of grief as an indication of madness. In *Zoonomia*, Darwin also categorized "melancholy insanities" as those which are "attended with despair and inaction" and suggested that the patient should be confined if he is either likely to injure himself, which "must be judged of by the despondency of his mind," or "cannot take care of his affairs" (2:350, 352). These are, of course, qualities that Seward emphasizes as being present in Darwin's son. Moreover, Darwin asserted that insanity is often hereditary (2:354). Thus, by detailing the more obviously categorizable qualities in Darwin's son and noting that Darwin himself called Erasmus "insane," Seward calls into question the sanity of the father as well.

Seward further details Darwin's lack of sensibility, both in relation to the loss of his son and in his human interactions in general. Writing of Darwin, she explains:

> The folly of suffering our imagination to dwell on past and irretrievable misfortunes, and of indulging fruitless grief, he often pointed out, and always censured. He relied much on self-discipline in that respect, and disdained, from deference to what he termed the prejudices of mankind, to display the outward semblance of unavailing sorrow, since he thought it wisdom to combat its reality. On occasions and subjects which he considered trivial, he professed to indulge human prejudice; but whenever, by mock assent, he extended that indulgence, a slight satiric laugh and a gay disdain lurking in his eye, counteracted the assumed coincidence. (409)

According to Darwin in *Zoonomia*, madness occurs when one loses touch with reality (2:350; 1:433). Here, Seward makes it clear that Darwin is the one who dwells outside of reality through an inability to feel and participate in expected human emotions. While Darwin often exemplifies his medical assertions through the case studies of his patients, Seward makes Darwin himself her case study as a means of mounting evidence for her own argument that it is not insanity to feel; rather, it is insanity *not* to feel.

However, it should also be remarked that Darwin's family took great offense at Seward's depiction of him as reacting callously or unemotionally to the loss of his son. In his 1879 biography of his grandfather, Charles Darwin treated Seward's portrayal as "an outrageously scandalous distortion of the truth about Darwin's character and his relationship with" his son, supposing that Seward's "malice" may owe to her feeling that "Darwin rejected her love," thus depicting her, as Teresa Barnard writes, as "a stereotypical scorned woman."[24] A few years after Erasmus Darwin's death, his daughter, Emma, wrote that Seward had dared "to accuse my dear Papa of want of affection and feeling towards his son," when in fact "he very frequently and *always with kindness* spoke of him," so that she concludes, "I want to scratch a pen over all the lies, and send the book back to Miss Seward ... and to swear the truth of what I have said before both houses of Parliament."[25] Following protests by both Robert Darwin and Richard Edgeworth, Seward published "a complete retraction in a number of magazines" regarding her portrayal of Darwin's indifference to his son.[26] In a personal letter to a correspondent discussing the *Memoirs*, Seward admits that "the late Dr. Darwin's family seem dissatisfied with my impartiality. I see they wanted to have had only the lights in his character shewn, and all its shades omitted. On the contrary, several of my friends murmur that I have not, as they term it, sufficiently stigmatized his irreligion," and she maintains that she presented her subject "with an even hand" (*Letters*, 6:136–37).

Significantly, while Seward presents Darwin and his poetry about plants as lacking sensation, she relates and endorses his scientific theory that plants themselves do possess the capacity to feel. Summarizing Darwin's "highly ingenious" *Phytologia* (1800), she writes, "Darwin's conviction that vegetables are remote links in the chain of sentient existence, often hinted in the notes to the *Botanic Garden*, is here avowed as a regular system," denoting that "plants have vital organization, sensation, and even volition," so that plants "are *themselves* capable of receiving comfort or discomfort" (*Memoirs*, 411–12). However, she also undercuts Darwin's success here by assuring that he "is neither the source, nor the first who drew the scattered hints of former philosophers concerning it, into a regular system," because Thomas Percival, in *Speculations on the Perceptive Power of Vegetables* (1785), "preceded him in maintaining that system from the press" (413). Thus, in Seward's discussion of "vegetable sensation," she emphasizes that while even plants in "the descending scale of existence" possess feelings, Darwin himself is not only defective in that capacity but also a plagiarist. Additionally, in a 1785 letter to Percival, congratulating him on his "ingenious pamphlet," Seward reveals that this notion of "conscious sensation" in vegetables "has long been a favorite hypothesis of mine," suggesting that she had this idea before either of these male writers (*Letters*, 1:17).

Indeed, even when relating the circumstances of Darwin's death, Seward continues to reinforce his inability to feel. She reports, "It was said that during his illness he reproved the sensibility and tears of Mrs. Darwin, and bid her remember that she was the wife of a philosopher," thus again indicating that Darwin not only lacked "sensibility" but also disapproved of it in others, even when the occasion warranted some display of emotion (*Memoirs*, 423). Moreover, in the midst of detailing Darwin's final hours, Seward embarrasses his memory further by repeating from other "imperfect sketches of his life" that "a strange habit was imputed to Dr. Darwin, which presents such an exterior of idiot-seeming indelicacy that the author of this tract is tempted to express her entire disbelief of its truth; viz. that his tongue was generally hanging out of his mouth as he walked along" (424–25). She states that she had "never witnessed a custom so indecent," but supposes that "the early loss of his teeth" may have "expose[d] the tongue to view while speaking" (425). Seward thus, under the guise of disputing this possible fault in Darwin, perpetuates it and undermines his claims to genius and knowledge in the popular imagination, making him appear mentally unstable and emotionally defective, as well as physically perhaps more worthy of laughter or pity than respect. Regarding the public reception of her *Memoirs*, Seward acknowledges that "my work on Darwin is likely to displease," for "the world of letters seem divided in two wide extremes; one half considering him as infinitely the first genius of his age, both as to poetic system" and "exalt[ing] him as having been almost superior to human frailty," and the other half "stigmatize him as an empiric in medicine, a Jacobin in politics; deceitful to those who trusted him, covetous of gain, and an alien to his God. What can I hope for, who spoke of him as he was" (*Letters*, 6:73–74). Insisting on the justice of her approach, in her literary naturalism Seward continued to incorporate her knowledge of botany and Darwin's medical theories on insanity not only in evaluating him and his own verses, but also those of other contemporary poets whom she likewise sometimes found to be deficient in sensibility.

Seward's Poppy, Cowper's "Crazy Kate," and Fuseli's Floral Madness

When discussing medical treatments for insanity in *Zoonomia*, Darwin frequently recommends doses of opium, a drug prepared from varieties of the poppy plant, or *Papaver somniferum*. Of the pains felt in diseases of irritation, which "are frequently succeeded by . . . madness," Darwin writes, "a much less quantity of opium will prevent them than is necessary to cure them" (2:122). Interestingly, he also often references similarities between opium and insanity, blurring the line between the cure and the disease. For example, as opium functions as a cure, Darwin remarks that insanity also sometimes seems to be the mind's or body's method of "curing" some other ailment within the organism (1:436). At the same time, in his section on the diseases of sensation, he compares the "delirium" of opium with that of certain kinds of madness, stating, "delirium is produced for a time by intoxicating drugs, as fermented liquors, or opium: a permanent delirium of this kind is sometimes induced by" pleasurable insanities (1:394). In *Zoonomia*, Darwin provides numerous case histories in which he exhibits the successful treatment of various illnesses with opium (1:438). King-Hele relates that Darwin prescribed opium to Tom Wedgwood, and that this may have influenced Coleridge in trying the drug, leading to both men's addictions, and even wonders whether "Coleridge might have come to regard Darwin as responsible," thus helping to explain "why Coleridge turned against Darwin in later years."[27] In Maria Logan's poem, "To Opium" (wr. 1788?, pub. 1793), she spends the final stanzas hoping that Darwin will include the poppy in his *The Loves of the Plants* and suggests in a note that she had seen drafts of that work before its publication.[28] In fact, Darwin does briefly treat the poppy in canto 2 of his poem, and discusses the production of opium in a scientific footnote there.

Seward's sonnet "To the Poppy" (1799) brings this flower and its effects into the realm of sensibility, versifying opium's medicinal ability to deaden pain and produce a sense of well-being or euphoria, as well as a kind of madness. Addressing the poppy itself, she presents the tale of a "love-craz'd Maid":

> While Summer Roses all their glory yield
> To crown the Votary of Love and Joy,
> Misfortune's Victim hails, with many a sigh,
> Thee, scarlet Poppy of the pathless field,
> Gaudy, yet wild and lone; no leaf to shield
> Thy flaccid vest, that, as the gale blows high,
> Flaps, and alternate folds around thy head.—
> So stands in the long grass a love-craz'd Maid,
> Smiling aghast; while stream to every wind
> Her garish ribbons, smear'd with dust and rain;
> But brain-sick visions cheat her tortur'd mind,
> And bring false peace. Thus, lulling grief and pain,
> Kind dreams oblivious from thy juice proceed,
> Thou flimsy, shewy, melancholy weed.[29]

Seward begins the poem with an image of social fulfillment, presenting "roses" as conventional symbols that "crown" those experiencing "love and joy." In contrast, she associates the poppy with melancholia and solitude as well as "misfortune's victim," inhabiting a "pathless field." Described as "gaudy," "scarlet," "wild and lone," the flower embodies qualities of the fallen-woman figure, unprotected from the ravaging of the elements and of mischance.[30] Likened to the poppy, the "love-craz'd Maid" embodies insanity in the form of her deranged smile and the filth and disarray of her clothing as she, too, is at the mercy of the environment. In the tradition of sensibility, the Maid represents an unspoken appeal because "oblivious" to her own pain and worthiness of sympathy. Since the Maid's thoughts are severed from the reality of her pain, the cognizant reader instead feels and seeks to understand her suffering, fulfilling the social expectation of compassion for those who cannot pity themselves. Eliciting sympathy for the Maid, Seward denigrates the poppy as "flimsy" and "showy," a mere "weed," whose medicinal relief brings only "false peace" and "brain-sick visions," indicating that the opium's "kind dreams" and oblivion for the Maid's "tortured mind" are "cheat[ing]" her with an escape that will not last. In Seward's portrayal, feminine forms thus act as both the deceiver (the poppy) and the deceived (the Maid), likening one to the other in a self-perpetuating comparison that entraps the Maid and gestures toward the negative consequences of opium's healing power, through which, again, the cure also functions as a disease.

Although Seward's "To the Poppy" provides no backstory for why the "love-craz'd Maid" is now "wild and lone," consciously or unconsciously, her poem appears to be influenced by the "Crazy Kate" episode in the first book of William Cowper's *The Task* (1785), "the most recognized eighteenth-century representation of female madness," which also has intriguing botanical associations.[31] Cowper's poem features a "serving maid" who fell in love with a sailor:

> She heard the doleful tidings of his death,
> And never smil'd again. And now she roams
> The dreary waste; there spends the livelong day,
> And there, unless when charity forbids,
> The livelong night. A tatter'd apron hides,
> Worn as a cloak, and hardly hides a gown
> More tatter'd still; and both but ill conceal
> A bosom heaved with never-ceasing sighs.

> She begs an idle pin of all she meets,
> And hoards them in her sleeve; but needful food,
> Though press'd with hunger oft, or comelier cloaths,
> Though pinch'd with cold, asks never.—Kate is craz'd. (1:545–56).³²

As I explore below, Seward expressed disdain for both Cowper and *The Task* in her published letters, and thus the resemblance between her "To the Poppy" and this depiction of "Crazy Kate" is striking and indicates that she may have viewed her sonnet as providing a corrective to Cowper through a more harrowing depiction of a "love-craz'd Maid"; for, while Cowper's Kate "never smil'd again" and always dwelled in realms of "fancy," Seward's Maid is "smiling aghast" and negligently "smear'd with dust and rain" while suffering from the more starkly termed "brain-sick visions" and a "tortur'd mind," as opposed to the poetically associated fancy (*Task*, 1:539).

Cowper, who titled the third book of *The Task* "The Garden," wrote two glowing reviews of Darwin's subsequent publications of *The Botanic Garden*, and his own verse depiction of "Crazy Kate" inspired further botanical artistic expression from the Britain-based Swiss artist Henry Fuseli.³³ Fuseli probably met Darwin in 1781, the same year this painter executed his most celebrated work, "The Nightmare," and in 1784 he also recommended the radical bookseller Joseph Johnson as an appropriate publisher for *The Botanic Garden*.³⁴ Fuseli contributed the design for "Flora Attired by the Elements" as the frontispiece for Darwin's *Economy of Vegetation*, as well as "The Fertilization of Egypt" and "The Tornado," which were both engraved by William Blake after Fuseli in that work, and an engraved version of "The Nightmare" appears in some editions of *The Loves of the Plants*.³⁵ Seward, too, personally knew Fuseli. As Martin Myrone recounts, in the summer of 1783 Brooke Boothby, a member, with Darwin, of the Lichfield Botanical Society, organized an elaborate Gothic pageant for Fuseli in which friends and family played supporting roles in "costumes as fairies, monsters, knights and ladies, singing and dancing," and "the climax of the evening" came when Seward "descended from a tree dressed as 'the Muse of Elegiac Verse'" for her elegies on the deaths of both Captain James Cook and Major John André.³⁶ Myrone explains that "rumors of [Fuseli] indulging in opium use and eating raw pork to fuel his sickly imagination lent a pathological element to his public reputation, making him one of the best known, and certainly most controversial, artists of the day."³⁷ One reviewer in the *Morning Post* called Fuseli a "genius

FIGURE 10. *Mad Kate*, Henry Fuseli, 1806/7. (Image courtesy of Frankfurter Goethe-Museum, © Ursula Edelmann—ARTOTHEK)

run mad," and Horace Walpole, on seeing an exhibition of Fuseli's paintings, simply noted in the margins of his catalogue, "shockingly mad, madder than ever; quite mad."[38]

Fuseli's contributions to Darwin's botanical publications arguably continued to shape his own work as his later painting, *Mad Kate* (1806–7), portrays that figure from Cowper's *The Task* with clearly floral features. Additional real-life and botanical influence may have come from Richard Polwhele's couplet-satire *The Unsex'd Females* (1798), in which Polwhele attacks Mary Wollstonecraft's advocacy of botany in young women's education and employs Darwin's commentary on sexual "adultery" among certain plants to reference her "manifest adultery" with the married Fuseli, as well as her affair with the unmarried Imlay, who deserted her, causing her to become suicidal and to attempt to drown herself.[39] In Fuseli's depiction, Cowper's Kate becomes a plant, resembling the *Arum maculatum* with common names including lords-and-ladies and jack-in-the pulpit, so that her wind-blown cloak forms the plant's spathe, while she herself embodies the plant's spadix. Fuseli hauntingly captures Kate's unsmiling "craz'd" glare and even the pins in her sleeve, begged from passersby as described in Cowper's poem, placing her next to the turbulent sea that is both symbolic of her mental and emotional disarray and associated with her lost love.

If Seward portrays Darwin as a kind of mad scientist figure devoid of sensations, she casts Cowper in the category of "mad poet," perceiving in his life and poetry similar disconnections from reality. Fascinatingly, in her correspondence with Robert Fellowes, with whom the critique that Darwin's poetry lacks "sensation" originated, she debates the existence of this same quality in Cowper's works. While Fellowes enthusiastically attributes "sensations" to *The Task*, Seward denies its presence there and instead accuses Cowper of "misanthropy" and "egotism," stating that "no composition can breathe more inward self-content" (*Letters*, 5:373, 328). According to Seward, rather than displaying sensibility's capacity for compassion and connection with other human beings, Cowper and his works are too solipsistic. She similarly disapproves of his Calvinism and blames this "gloomy sect" for his "horrid ... insanity" (5:375–76). In 1799, she assures Fellowes that Christianity, in line with concepts of rational sensibility, represents "rational faith" and "a system of consistent justice, mercy, benevolence, and happiness," while the "Calvinistic school" instead espouses "mystical tenets" that transfer "the voluntary scourge, and all the dark train of monkish self-inflictions, from the body to the mind," and thus pose a threat both to mental health and to a true understanding of God (5:246–47).[40] For Seward, writing to William Hayley in 1803, Cowper's Calvinism conflates with mysticism, insanity, egotism, and an antisocial lack of literary patriotism.[41] She writes of Cowper, "Whether from the narrow and miserable principles of Calvinism with which he was so deeply tinctured ... or whether from a native taint of insanity, I know not, but I see him, with all his inherent good properties, a vapourish egotist" in "the mystical and dull jargon of his letters," for "all is egotism" and "absorbed in himself, he appears to have been ignobly inattentive to the works of poetic genius which have adorned his country from Milton's time to the present" (6:60, 63). Seward's concern with Cowper's "eternal sallies of egotism and self-importance" merits further consideration as she does not find all egotism so off-putting, and defends Hayley and William

Mason, among others, for exhibiting this quality (6:162, 2:206). Further, she does not limit its religious association to Calvinism, but additionally warns of its use in orthodox faith as well, confirming that "nothing is more disgusting to me, and indeed to the generality of people, than dictatorial egotism from the pulpit" (5:348). Seward gives insight into this distinction when she responds to a letter from Walter Scott where he fears expressing egotism in discussing his own work, and she comforts him, stating, "The egotism of mean minds disgusts, that of elevated ones is interesting. Rousseau was the greatest egotist existing, yet lost no reputation by that excess" (6:40). In Seward's use, then, egotism becomes a matter of aesthetic judgment, based on the quality and sensibility of the literary mind employing it and its affect on the reader. Interestingly, another poet whose botanical displays of egotism and mysticism "disgust[ed]" Seward, drawing her accusations of insanity, was William Wordsworth.

Wordsworth's "Daft" Daffodils

According to Seward, Wordsworth exemplifies poetic insanity especially in his treatment of flowers in his poem "I Wandered Lonely as a Cloud" (wr. 1804, pub. 1807). She writes in a published letter, "Surely Wordsworth must be mad as was ever the poet Lee," thus comparing him with Nathaniel Lee, a Restoration-era playwright who spent time confined in Bedlam (6:366). Seward admits feeling "contemptuous astonishment and disgust" while reading

> about [Wordsworth's] dancing daffodils, ten thousand, as he says, in high dance in the breeze beside the river, whose waves dance with them, and the poet's heart, we are told, danced too. Then he proceeds to say, that in the hours of pensive or of pained contemplation, these same capering flowers flash on his memory, and his heart, losing its cares, dances with them again.
> Surely if his worst foe had chosen to caricature this egotistic manufacturer of metaphysic importance upon trivial themes, he could not have done it more effectually! (6:366–67).

Although many elements are at play in Seward's critique, by stating that these verses make Wordsworth appear "mad," she also cleverly puns on his choice of flower. At this time, the daffodil was sometimes termed as "daffodilly" or "daffy," the last of which was synonymous with "daft," meaning "of unsound mind, crazy, insane, mad."[42] Thus, in Seward's analysis, through its very etymology, this floral affiliation and Wordsworth's identification with the "dancing daffodils" suggests his own madness. Moreover, as Seward doubtless knew, the daffodil belongs to the genus *Narcissus*, displaying additional wit in her accusation regarding Wordsworth's "egotis[m]." She thus anticipates the famous mockery of this poet's egotism from other contemporaries. William Hazlitt, for instance, writes that Wordsworth's "egotism is in some respects a *madness*; for he scorns even the admiration of himself, thinking it a presumption in any one to suppose that he has taste or sense enough to understand him," and another reviewer observes "how ludicrously he overrates his own powers. This we all do; but Wordsworth's pride is like that of a straw-crowned king in Bedlam."[43] And, as Noel Jackson relates, in the context of the era's aesthetic and scientific writing, Wordsworth may have had cause to dread judgment of his attachment to sensations of the past not only "due to his awareness of how such attachments could suggest the poet's indolence or self-absorption, but just as plausibly from an apprehension of how closely such experiences resemble a kind of madness."[44] In regard to egotism, Wordsworth perhaps knew of the daffodil's association with the Narcissus myth, enhancing the daffodils' placement beside the water. Describing this poem as a satire on the poet himself, a "caricature" that places "metaphysic importance upon trivial themes," Seward also anticipates Coleridge's criticism of these verses in *Biographia Literaria*, where he similarly asserts that the verses contain "thoughts and images too great for the subject" in "what might be called *mental* bombast."[45]

Nevertheless, modern scholars suggest additional scientific meanings and influences for Wordsworth's poem. Marjorie Levinson, for example, posits that Wordsworth's physical and psychological wandering in these verses may be inspired by both Luke Howard's scientific *Essay on the Modifications of Clouds* (1803) and Baruch Spinoza's theories of nature and affect, so that Lisa Ottum and Seth T. Reno view the poet as representing an "affective state compatible with an ecological sensibility."[46] King-Hele asserts that Darwin's ideas about plants became "an essential part of [Wordsworth's] basic belief in Nature."[47] And, as Fred Blick has shown, when Wordsworth states that the daffodils "flash upon that inward eye" and compares them with the "twinkle" of stars and the "sparkling waves," he may be drawing on Darwin's discussion of the "Elizabeth Linnaeus phenomenon."[48] Elizabeth Linnaeus, daughter of the famous Swedish botanist, published a paper in the *Acts* of the Royal Swedish Academy of Science in autumn 1762 describing her observations of flashes of light being emitted from yellow Indian Cress flowers. Other naturalists, including Lars Christian Haggren, confirmed this discovery, and news of it was published in British journals. Darwin included notes regarding these flashing flowers in *The Loves of the Plants*, and Coleridge appended an endnote to his poem "Lines Written at Shurton Bars, Near Bridgewater, September 1795," which "reproduced almost verbatim . . . the note from Darwin's" work; however, Coleridge omitted this note after 1803.[49] As is often remarked, Wordsworth's "I Wandered Lonely as a Cloud" originated with Dorothy Wordsworth's journal entry that details a sighting of daffodils at Ullswater on 15 April 1802, and it was William's wife, Mary, who provided the lines "They flash upon that inward eye / Which is the bliss of solitude."[50] In fact, M. M. Mahood shows that the women in Wordsworth's life demonstrated more interest and knowledge in the science of botany while Wordsworth, by his own admission, neglected the subject, "into which neglect I was partly seduced by having in that early part of my life, my sister almost always at my side in my wanderings, and she for our native plants was an excellent botanist."[51] Still, Seward disapproves of the daffodil poem which, in her view, signals Wordsworth's poetic madness through his egotism and thus failure to appropriately express sensibility, a shortcoming that she attributes to his poetic theories more generally.

Examining Wordsworth's poetic doctrine, Seward voices concern about him as a model for younger poets, declaring that he "has undoubtedly genius . . . but on the whole, it is not first rate" and pronounces his poetry inferior to that of both Coleridge and Southey (*Letters*, 6:258, 280, 367). She references the preface to *Lyrical Ballads*, explaining that Wordsworth "is right in observing that the use of common life simple language in verse is frequently a beauty, but not right in extending that use to all modes of phraseology within the limits of the immodest, the disgusting, and the ungrammatic" (6:258). This results in his poetry being both "meanly familiar" and "obscure" (6:258). For Seward, Wordsworth's poetic "system" therefore risks producing verse that is "vague, indiscriminate, and dangerous" because often lacking "polished elegance" and "delicacy" (6:260–61). Further displaying what Seward found so "dangerous" about Wordsworth's poetry, she cites another poem where he demonstrates, in her view, "successful and unsuccessful attempts at sublimity" and "goes rumbling down the dark profound of mysticism, whither my comprehension strives to follow him in vain" (6:367). In Wordsworth's "attempts" to reveal something "importan[t]" and "sublim[e]" in the "trivial," he aligns himself, according to Seward, with "mysticism" and the "metaphysic[al]," making him appear mad due to the intensely egotistical experiences related in his poems that become inaccessible or "obscure" due to their extreme subjectivity. Describing this affinity between madness and egotism, Frederick Burwick, paraphrasing Foucault, writes, "Art is twin-born with madness . . . yet the two grow apart because madness is self-centered, whereas art strives to communicate."[52] Adela Pinch observes that Wordsworth's "excessive investment in his own emotional responses has led somewhat paradoxically to the assessment of his character . . . as cold, inhuman, indifferent to the suffering of others."[53] Regarding this poet's "disproportionate acts of attention and emotion," Geoffrey Hartman similarly asserts that Wordsworth is "inevitably linked to a self-consciousness that may seem egotistical."[54] In Seward's view, the isolating egotism of Wordsworth's poetic madness ensures his art's failure to communicate its meaning outside the self, which indicates a failure of connection, of sensation or sensibility.

Of course, John Keats, too, would later accuse Wordsworth of representing the "egotistical sublime" and perhaps provides an unspecified critique of the older poet's daffodil poem, exemplified through another floral analogy. He writes: "Poetry should be great and unobtrusive, a thing which enters into one's soul and does not startle it or amaze it with itself but with its subject.—How beautiful are the retired flowers! how they would lose their beauty were [they] to throng into the highway crying out, 'admire I am a violet! dote upon me I am a primrose!'"[55] For Keats, Wordsworth's narcissistic intrusions thus arguably corrupt flowers' "great and unobtrusive" poetic beauty, the divine artistry that "enters into one's soul." In this way, both Seward's and Keats's critiques of Wordsworth's egotism, mysticism, and "attempts at sublimity" could be said to echo Darwin's description in *Zoonomia* of "pleasurable insanities," involving hallucinations and delusions categorized as "mania" (2:356). According to Darwin, in those instances that produce "pleasurable sensations, as in personal vanity or religious enthusiasm; it is almost a pity to snatch them from their fool's paradise, and reduce them again to the common lot of humanity; lest they should complain of their cure" (2:354). Wordsworth was familiar with Darwin's *Zoonomia* and drew on it for "psychological and associationist context" in the *Lyrical Ballads* and elsewhere.[56] For Seward, Wordsworth's daffodils "in high dance" present an "astonish[ing]" instance of such madness and detachment from reality, placing the poet in danger of solipsistically failing to connect with his fellow human beings, which also marks his failed sensibility.

Clare's "Dark" Botany and Mystical Madness

Although Seward never considered Darwin's poetry "as the best model for rising genius to adopt" and declared Wordsworth's "system" of verse "dangerous" to younger poets, the Romantic-era poet who is perhaps most frequently connected with insanity and precise natural description, John Clare, deemed both Darwin and Wordsworth important poetic influences. Seward died over a decade before Clare's first volume of poetry was published in 1820, but he provides useful insights into her legacy and the evolving relationship between literature, botany, and madness. Whereas Seward attacked Charlotte Smith for departing from the "legitimate" Miltonic sonnet form, Clare was first inspired to write sonnets after reading some of those by Smith, and he viewed Milton's sonnets as "too rule-bound."[57] While Seward deprecated the "madness" of Wordsworth's mysticism, Clare, whom scholars often describe in terms of mysticism within a "religion of nature," found method in Wordsworth's madness, praising him as a poet who "defies all art and in all lunatic Enthuseism of nature he negligently sets down his thoughts from the tongue of his inspirer" (Bate, 253, 187). Yet, like Seward, Clare ridiculed Wordsworth's "nursery rhyme" manner as being easily parodied (188). Admitting that he thought Wordsworth's poems sometimes contained "affected fooleries," Clare appreciated that "still with his faults and abilities he is a poet with whom for origionallity of description the present day has few if any equals—for the present," suggesting, according to a recent biographer, that Clare harbored hopes of soon surpassing Wordsworth in originality of description (187). Nevertheless, Clare also received influence from Darwin's verse; as late as 1818, a reviewer admitted that, "in the powers of description, [Darwin] has few superiors within the range of British poetry," and both Clare's contemporaries and modern literary criticism sometimes fault this younger poet as being too "strictly a descriptive poet."[58] As Hugh Haughton puts it, "Romantic poetry is fascinated by the poetry of 'vision', Clare's by *seeing*—something very different."[59] This emphasis on the eye, on "seeing" and visual objects, thus arguably aligns Clare with Darwin's poetic theories and methods of the "eye" in ways that may further help to explain Clare's relative historical neglect in relation to the Romantic canon.

At the same time, while Clare possessed much local knowledge of plants and frequently fills his poetry with a sense of immediacy and the close attention to nature that one might find in a naturalist's observations, unlike Darwin he generally avoids using the technical language of Linnaean taxonomy in his verse. In his first poem, "The Morning Walk," probably written in 1806, when he was thirteen, Clare lists the flowers he sees and mentions Darwin by name in a later-revised stanza:

> Nay, had I Darwin's prying thought,
> Or all the learning [John] Ray has taught,
> How soon description would exhaust
> And in sweet Flora's lap be lost.[60]

Clare wrote that John Hill's *Family Herbal* (1754) "gave me a taste for wild flowers which I loved to hunt after and collect to plant in my garden which my father let me have in one corner of the garden" (Bate, 102). He also owned a copy of James Lee's *Introduction to Botany* (1760), which explained Linnaeus's sexual system of plant classification and binomial nomenclature, and he admits that this book briefly inspired him to collect plants "into familys and tribes but it was a dark system & I abandoned it with dissatisfaction" (103).[61] He complains,

> I have puzzled wasted hours over Lee's Botany to understand a shadow of the system so as to be able to class the wild flowers peculiar to my own neighborhood for I find it would require a second Adam to find names for them in my way & a second Solomon to understand them in Linnaeus's system—moder[n] works are so mystified by systematic symbols that one cannot understand them till the wrong end of ones lifetime & when one turns to the works of Ray, Parkinson, & Gerard where there is more of nature & less of art it is like meeting the fresh air & balmy summer of a dewey morning after the troubled dreams of a nightmare.[62]

Fascinatingly, in Clare's description, the Linnaean system of botanical classification becomes a kind of madness through concerns reminiscent of those voiced by Seward when critiquing other male poets so that, for Clare, Linnaeus's technical language makes botany appear "mystified," obscure, inaccessible, esoteric, and thus egotistical in the sense of exclusionary; for him, despite its objective to provide a universal system, it instead frustrates communication and social connection, arguably resulting in a "nightmare" of failed sensibility.

According to Clare, Linnaean botany, that "hard nickname-y system of unutterable words now in vogue, only overloads [its subject] in mystery."[63] He laments, "I love to look on nature with a poetic feeling which magnifys the pleasure . . . yet naturalists & botanists seem to have no taste for this practical feeling[.] they merely make collections of dryd specimens classing them after Leanius [Linnaeus] into tribes & familys & there they delight to show them as a sort of ambitious fame."[64] For Clare, Linnaean "naturalists & botanists" too often lack "poetic feeling"; however, he appears to except Darwin from this accusation, perhaps because that older poet emphasizes plants' lives, rather than their deaths in "collections of dryd specimens." Indeed, one of Seward's poetic protégés and most frequent correspondents, Henry F. Cary, who wrote a long essay, "On the Life and Writings of Erasmus Darwin," published in the London Magazine in December 1822, became friends with Clare; in a letter to Cary in 1827, Clare wrote, "I have latterly read a good deal of Darwin's poetry & I am astonished to think why he is neglected so much for I am sure there is passages of uncommon harmony & beauty in them."[65] At the same time, Clare appreciated plants within their natural environments and according to their local or common names, apart from scientific efforts to impose alternate identities and meanings. In fact, Mahood notes with precision that Clare names 370 plants in his poetry and prose.[66] While Seward poetically referred to the poppy as a "melancholy weed," Clare enthused, "I love all wild flowers (none are weeds with me) affectionately—there is a little white starry flower with pale green grassy leaves grows by woodsides and among bushes—I know not its name but it is a boyish favorite and the same that those stanzas address as 'a nameless flower obscurely blooming in a lonely wild'" (Bate, 151). Clare here means his 1820 poem "To an Insignificant Flower," but his misremembering of it as a "nameless flower" emphasizes his detachment from particular or consistent naming of natural objects, as when "woodbine sometimes figures as 'honeysuckle', bindweed as 'bell flowers' and ragged robin as 'meadow pink,'" or when he poignantly writes in another poem, "hail to the nameless."[67]

Clare's empathy for plants and frequent resistance to naming them within Linnaean binomial nomenclature, his poetic exclamations celebrating natural objects' fluid identities ("hail to the nameless"), and his habit of calling a single species of flower by various common names finds an interesting parallel during his years at the asylum in Northampton, where an editor of a local newspaper visited him, reporting that Clare claimed lines from both Shakespeare and Byron as his own. The editor records, "'Yours!' I exclaimed, 'Who are you? These are Byron's and Shakespeare's verses, not yours!' 'It's all the same' he answered, changing a quid from one cheek to the other, 'I'm John Clare now. I was Byron and Shakespeare formerly. At different times you know I'm different people—that is the same person with different names'" (Bate, 474). If Clare's statement does blur the line between genius and madness, its gesture toward interconnectedness also reinforces his philosophy regarding the fluidity of naming and identities ("the same person with different names")—here, for poets as well as plants.

Clare often suggests that he identifies with the natural objects that he describes in his poetry, and thus perhaps part of his resistance to classifying them is due to his own awareness of being classified himself, first as a "peasant poet" due to his social class, and then as a "mad poet" as others sought to place him within various categories of insanity. Jonathan Bate relates that Clare's experiences of psychosis, especially auditory and visual hallucinations, encouraged posthumous diagnoses of schizophrenia in the early twentieth century, but that manic depression or "bipolar affective disorder" seems more likely to have been his illness (410, 412). In a medical perspective closer to Clare's own time, Darwin's *Zoonomia* pathologizes numerous experiences that can be found in this younger poet's life and verse, and Clare's own doctor at the private asylum in Epping Forest, Matthew Allen, wrote an *Essay on the Classification of the Insane* (1837).[68] Indeed, when Clare was admitted to the Northampton General Lunatic Asylum in 1851, his doctor suggested on the admission papers that his insanity was brought on "after years addicted to poetical prosing" (Bate, 466). As Whitehead puts it, poetry "had a dual identity, being granted privileged insight into madness, yet uniquely correlated to going mad, or demonstrating that one had gone mad."[69] Although Clare earlier called Linnaean botany a "dark system" and arguably associated it with madness through his sense of its inaccessibility and alienating obfuscation of its subject and the individual natural object, creating a lack of connection and sensibility, in his asylum poetry he hauntingly applies such adjectives to himself, expressing his own mental anguish and "bouts of dark depression": "My Mind Is Dark and Fathomless And Wears / The Hues of Hopeless Agony," so that "Day seems my night and night seems blackest hell."[70] Additionally, while he previously condemned naturalists and botanists for lacking "poetic feeling" toward plants and sacrificing them as specimens for scientific examination and display in glass cases, in the asylum he expresses his own "isolation and numbness," stating now, "Nature to me seems dead & her very pulse seems frozen to an icicle in the summer sun."[71] If Clare's poetic genius and madness had enabled him to identify with the more lively and dynamic aspects of plant life, he likewise mirrors the *in*sensibility involved in their mistreatment when made to feel like a taxonomized specimen himself.

Seward, Mental Health, and a Botanical Bright Side

Seward's preoccupation with other poets' expressions of madness during the final years of her life may reflect her concern regarding her own mental health. At least as early as 1793, Seward complained of a mysterious illness, writing, "My sensations taught me to apprehend the slow accumulation of water in my brain," yet her "disease, whatever it is, has been wonderfully abated by the course of medicines prescribed by Dr. Darwin and Dr. Jones in consultation" (*Letters*, 3:215). In 1797, she reports having an internal "hemorrhage" so that she awoke "with streams of blood running down my throat, and threatening suffocation" (5:21). By 1799, she laments in a letter to a correspondent that "this oppressive malady in my head" prevents her from making progress in writing poetry (5:269). In 1804, the year she published the *Memoirs*, she complains that "my head is not steady" due to "the dangerous paroxysm," "the dreadful seizure in my head" (6:199, 193). A letter to Walter Scott in 1805 provides further insight into her declining condition, especially following Saville's death, "which tore from me health, and peace, and every earthly hope" causing an increased "dizziness of head" and

> sudden paroxysms, in which all the surrounding objects seem falling into chaos. These paroxysms are brought on by every attempt to stoop my head to read or write with any continuance. Thus am I obliged to employ an amanuensis for my letters, and to procure a friend to read to me audibly whatever I wish to peruse; nor can I sustain, without danger, a continued attention, and still less a chain of intense thinking. By this strange mysterious malady, which medicine has tried to combat in vain, the remnant of my days is destined to a gloomy suspension of every intellectual industry. (6:207–8)

Thus, Seward, who had always taken pride in her skills as a literary critic, or "literary naturalist," confidently ordering both poets and their works into classes that register early formulations of a literary canon, here describes a mental state of disorder in which "the surrounding objects seem falling into chaos." Marilyn Gaull notes that Linnaeus himself in old age, "for all the rationality of his system, lost his reason: having identified countless rare specimens, he could not recognize his own book; having named so much of creation, he could not even remember his own name."[72]

Interestingly, Seward's malady further connects her with sensibility. In 1806, three years before her death, Seward wrote to another friend, "You congratulate me in your conviction that the now long-enduring malady in my head has in no degree impaired my intellects.—Perhaps not, except that it has considerably affected my memory. Names escape me strangely.... These are trifling privations, only as they threaten greater" (6:268). That last sentence seems key as Seward seeks to reassure herself that her symptoms—dizziness, difficulties with memory, and so on—do not necessarily indicate serious issues and are worrisome "only as they threaten greater" concerns developing in the future. At the same time, Seward's frequent communications to friends regarding her illness recall sensibility's paradox. For, while she challenged medical theories aligning excessive sensibility with disease and madness, and her complaints of suffering may embody a potential for solipsism of which she accused Cowper and Wordsworth, her illness also places her in sensibility's sociable and socially acceptable position of eliciting others' sympathy or compassion.[73] In the years preceding her death, Seward suffered from rheumatism, a breast injury, and a broken kneecap, as well as these "occasional hemorrhages, dizziness, and violent headaches" but, through all of this, as Barnard phrases it, she continued to write "compulsively until the pen slipped from her hand" when she died in 1809.[74]

Seward's personal concerns about mental health also arguably shed new light on her vindication of Mary Delany and this sociable bluestocking's celebrated artistry in paper-flower collages or "paper mosaiks" in the *Memoirs*, asserting women's continued capacity for intellectual and creative brilliance and achievement into the late years of their lives. Repudiating Darwin's depiction in *The Botanic Garden* of Delany's "miraculous Hortus Siccus," completed in 1783, Seward declares that this "forms one of the most censurable passages in the whole poem" because "entirely false" (*Memoirs*, 315). She admiringly explains that in Delany's "representation of plants and flowers, native and exotic, and which fill ten immense folio volumes," she "used neither the wax, moss, or wire attributed to her" by Darwin and in fact "employed no material but paper, which she herself,

FIGURE 11. *Pancratium maritimum*, or sea daffodil, by Mary Delany, 1775. In her famed collection of 985 "paper mosaiks," *Flora Delanica*, Delany labeled the plants' Linnaean and common names. (© The Trustees of the British Museum)

from her knowledge of chemistry, was enabled to dye of all hues, and in every shade of each; no implement but her scissors, not once her pencil; yet never did painting present a more exact representation of flowers of every color, size, and cultivation" (315–16). Moreover, Darwin "misrepresents [Delany] as being assisted," but Seward clarifies that "she had no assistant; no hands, but her own, formed one leaf or flower of the ten volumes" (316). Relating this incident, Seward again questions Darwin's lack of sensibility or feeling, for while Delany's "family were mortified by a description which they justly thought degraded her peculiar art" and communicated this to Darwin, he "did not cho[o]se to alter the text" in his poem's subsequent editions (316–17). Thus, whereas Darwin diminishes the accomplishment and originality of Delany's work, Seward extols this female botanical artist, calling her a "prodigy of female genius" (*Letters*, 3:195–96). Significantly, Seward emphasizes that Delany "began this her astonishing self-invented work at the age of seventy-four" and marvels, "What a lesson of exertion does the invention and completion of such a work, after [seventy], give to that hopeless languor, which people are so prone to indulge in the decline of life?" (*Memoirs*, 316; *Letters*, 3:196). In this way, although Seward began to fear that "the remnant of my days is destined to a gloomy suspension of every intellectual industry," she defends Delany's botanical example as a hope for what is creatively possible even as time takes its toll on the body and mind.

As I have shown in this chapter, according to Seward, botanically affiliated poems by Darwin, Cowper, and Wordsworth highlight each of these male authors' particular forms of "madness" through their shortcomings in sensibility, an accusation that Clare sometimes made against Linnaean botanical studies more generally. Seward's imputations of madness as an effort to classify and control the success of other poets' works participate in what became a common tactic in nineteenth-century literary criticism. As James Whitehead writes, "The Romantic mad poet primarily emerged or was constructed in the nineteenth century . . . from criticism, biography, and other discourse built around and over the remains of Romantic writers."[75] Indeed, each of the now-canonical male Romantic poets was attacked as "mad" by critics and reviewers of their poetry, and it was the periodical reviewers who turned "the link between madness and poetry into a widely popular and enduring cultural mythology."[76] In contrast, Porter suggests that "the Romantics as a whole saw the writer not as psychologically peculiar but as truly healthy."[77] According to De Quincey, Wordsworth and Coleridge maintained "that mad people are the dullest and most wearisome of all people," thus implying their own sanity through the distanced critique.[78] Within this era coincided popular conceptions of both the "mad scientist" and "mad poet" figures, setting the stage for each of these emerging professions of science and literature to sometimes accuse the other of madness, while each also claimed for themselves the high ground of stability and rationality.[79] As I explore in the conclusion, De Quincey, too, employed his authority as a literary critic to impute madness to particular kinds of literature. However, while Seward attributed madness to certain works or authors in her literary criticism to encourage (at least theoretically) sociability and connection, De Quincey does so to promote Wordsworthian principles of masculinization and separation in the context of literature's relationship to science. Analyzing several of De Quincey's essays that provide insight into his famed division of the literature of knowledge from the literature of power, I display the larger literary and sociopolitical consequences of his attacks on scientific literature as he and other critics and reviewers during the nineteenth century established now-canonical conceptions of British Romanticism.

Conclusion

De Quincey, Hazlitt, Wordsworth, and the Critical Fate of Romanticism and Scientific Literature

As I discussed in the introduction, by the 1820s and 1830s a number of British botanists, scientific societies, and universities had begun shifting away from Linnaeus's "artificial" system of plant taxonomy and toward the "natural system," advocating a more comprehensive approach to plants' morphological features and aligning with the classifications of Continental botanists such as Jussieu and Candolle. However, I wish to stress that this change in botanical practice should not necessarily be viewed as a primary reason for the dissolution or transformation of the era's movement of scientific literature. Many authors and advocates of scientific literature, including John Aikin, Charlotte Smith, Erasmus Darwin, and Anna Seward, earlier challenged aspects of the Linnaean system and enacted negotiations between literature and science. These writers thereby provided the possibility for methods of collaboration and mutual or interconnected improvements within these developing disciplines. Yet this potential in many ways ceased or altered when principal figures responsible for the professionalizations of science and literature, such as Humphry Davy and William Wordsworth, emphasized instead these fields' competition with one another and contended "for the same social niche and for the attention and acclaim of the same public."[1] As I argued in my previous book, *Questioning Nature*, while both the natural sciences and scientific literature often practiced intertextuality and embraced collective, collaborative approaches in the late eighteenth and early nineteenth centuries, these methods shifted with the developing, separate professionalizations of science and literature as the nineteenth century progressed. Although the works of Wordsworth and other now-canonical male Romantics of course display intertextuality, these writers also sometimes contributed to the aesthetic fantasy that their writing achieved unindebted originality, and critics and reviewers during the 1820s and 1830s retroactively simplified and mythified this conception of the male Romantic poet as a "solitary genius" producing works of autonomous originality.[2]

In contrast to this idealized portrayal of some of the now-canonical male Romantic-era writers, throughout the nineteenth century male literary critics, reviewers, and scientists instead often theorized women's capacities in ways that disparaged or precluded women's abilities to achieve imagination (which, as I discuss below, medical theorists complexly connected with sensibility) as well as originality. Numerous essays and treatises composed by nineteenth-century biologists, physicians, anthropologists, psychologists, and other scientists sought to provide conclusive evidence of female inferiority, especially based on their supposed lack of originality.[3] Likewise, while sensibility sometimes could be perceived as distinguishing intellectual or moral superiority, this quality, especially in its association with women, became medically affiliated with forms of disease, melancholia, and insanity when "in excess."[4] At the same time, many nineteenth-century European scientists additionally sought evidence to support claims of European intellectual superiority, so that scientific or biological sexism and racism often pursued similar lines of argument, appropriating "ideal" expressions of both sensibility and originality for European men.

As I have stated, for both male and female writers during the Romantic era, the melding of natural history and literature could be a means to literary authority and originality. However, women writers participating in this movement of scientific literature, and especially poetry, often experienced a kind of dual discrimination from male critics and reviewers during the nineteenth century. In addition to the growing critical disfavor toward scientific literature more generally as a means of denigrating its obvious intertextuality, male critics and reviewers in this era often imposed gender-specific expectations on women writers, holding their literature to a different standard of modesty, sentiment, tone, and propriety that sometimes affected reviews disparaging women's scientific engagements. Correspondingly, it became less socially appropriate or acceptable for women to versify scientific subjects as the natural sciences became more secularized through increasing challenges to biblical tradition around the 1820s and 1830s and subsequent decades. In contrast to earlier expectations of natural theology that had enabled imaginative authors, especially women, to justify blending literature and technical science as a pious pursuit, science's greater secularization encouraged many women poets of the late Romantic and Victorian eras instead to turn more fully toward religious themes within their verse. Simultaneously, the separate professionalizations of science and literature, imposing a greater divide between these disciplines as science retreated into academies and increasingly excluded the participation of nonspecialists, made it more difficult for both men and women writers of imaginative literature to feel empowered to provide new or technical scientific information through their literary works.

In this conclusion I argue that, in the early nineteenth century, William Wordsworth, William Hazlitt, and especially Thomas De Quincey not only portrayed literature as competing with the sciences for cultural authority but also paradoxically created a mirroring effect between these disciplines, so that "recognizable" literature through conceptions of the so-called "literature of power" embraced prejudicial aspects of "professional" science. In this way, these male writers thereby furthered efforts to disparage and exclude from the literary canon especially women (often in connection with writing scientific literature) and (for De Quincey) non-Europeans by reinforcing "scientific" claims of inferiority based on sex and/or race. Moreover, these male writers' appropriations and manipulations of conceptions of originality and sensibility crucially enacted this exclusionary shift in standards of literature. As each of these men's writings became extremely influential in forming the literary canon, I examine pertinent moments in several texts, including Hazlitt's *Lectures on the English Poets* (1818), select letters from Wordsworth, and De Quincey's essays on knowledge and power (1823, 1848) as well as some of his assertions on women and race. In chapter 6, I demonstrated ways in which madness was sometimes associated with both an excess and dearth of sensibility and could be used by literary critics in attacks against particular writers or approaches to literature; here, I show, for instance, that De Quincey deployed this strategy against scientific poetry and suggested that the melding of science and literature represented a kind of madness, even as he adopts scientific theorizations that allow him to exclude literary authors based on sex and race.

In fact, the scientific literature that I have analyzed in this study is imperative to understanding the process of both literature's and science's increasing professionalizations around the 1820s and 1830s. This is because key members of these disciplines separately made efforts to masculinize public perceptions of their fields after science and literature became jointly associated with women, largely through this movement of scientific literature that encouraged women's study of the natural sciences. The British Association for the Advancement of Science (BAAS), for example, which formed in 1831 for the purpose of professionalizing science, "chose not to include the arts within its ambit, an exclusion that effectively institutionalizes the separation of poetry and science at the inception of British professional science," and represents just one of this era's burgeoning scientific organizations that also barred women's participation.[5] However, perhaps more disconcerting is that, unlike science's supposedly "objective" assertions, some important male writers and literary critics during the first half of the nineteenth century, including Wordsworth and De Quincey, espoused literary sexism and/or racism while claiming to champion the "heart." In other words, although professing morality (or even spirituality), sensibility, and compassion, certain nineteenth-century male critics, authors, and reviewers instead not only appropriated and altered earlier understandings of sensibility but also employed this concept as a means of justifying their standardization of strategic and selective exclusions, especially regarding women and non-Europeans. Thus, while purporting to reject science, they actually adopted some of its evolving prejudicial principles of biological sexism and racism in key texts of Romantic-era aesthetics that also shaped the literary canon. In particular, De Quincey's definition of the "literature of knowledge" (aligning knowledge with science) in opposition to the "literature of power" has been recognized as "a distinction of immense historical effect," helping "to shape the nascent concept of English literature as something inherently separate from other kinds of writing" and to "justify the invention of a tradition and a canon and thus foster[ing] literature's emergence as an academic discipline and a school subject."[6] I thus argue that Romanticism and the literary canon as we know it (and, in many ways, the scientific disciplines themselves), were largely formed in opposition to this late eighteenth- and early nineteenth-century movement of scientific literature, which had become so strongly associated with women writers.

Critical Accountability: Recognizing Romanticism's Abuse of "Power"

Although several of Thomas De Quincey's works form the main focus of this conclusion, it is useful to briefly acknowledge the earlier role of William Hazlitt's influential text of literary criticism *Lectures on the English*

Poets. In *Lectures*, Hazlitt endorses and perpetuates Wordsworth's (and some other now-canonical male Romantic-era writers') division of literature from science, as well as this poet's efforts to appropriate sensibility and masculinize literature and especially poetry. Hazlitt describes Wordsworth as a poet who "hates all science and all art; he hates chemistry, he hates conchology [the study of shells]; he hates Voltaire; he hates Sir Isaac Newton ... yet he would be thought to understand them; he hates prose; he hates all poetry but his own."[7] Yet he deems Wordsworth "the most original poet now living," explaining that "his poetry is not external, but internal; it does not depend upon tradition, or story, or old song; he furnishes it from his own mind, and is his own subject. He is the poet of mere sentiment," and his poems "open a finer and deeper vein of thought and feeling than any poet in modern times has done, or attempted."[8] If Wordsworth, Seward, and other contemporary writers attacked Darwin for being "all eyes" and no heart, then, according to Hazlitt, Wordsworth has his own vision problems due to his "egotism," as "he sees nothing but himself and the universe."[9] Nevertheless, Hazlitt's assurance of Wordsworth's superiority to "any poet in modern times" works to condone and encourage this poet's principles, including his theoretical eradication of serious engagement with the natural sciences from verse, helping to ensure that his methods would become a standard guide for subsequent decades of literature.

In asserting Wordsworth's literary dominance, Hazlitt also restates this poet's critiques of Erasmus Darwin. Echoing Wordsworth's emphasis on great poetry as "permanent" and "immortal," Hazlitt writes that the best poets strive "to attain the highest excellence, sanctioned by the highest authority—that of time" and thus "are not afraid that truth and nature will ever wear out; will lose their gloss with novelty, or their effect with fashion."[10] He poignantly declares, "I have myself out-lived one generation of favorite poets, the Darwins, the Hayleys, the Sewards. Who reads them now?"[11] Conflating Darwin and Seward, and associating both with ephemerality, Hazlitt repeats Mathias's, Wordsworth's, Coleridge's, and others' earlier attacks on Darwin's poetry as merely representing "novelty" and "fashion"; he also perpetuates the lumping together of certain poets as belonging to "the Darwinian school" ("the Darwins"), regardless of their actual originality or individual and distinct styles and goals, placing these "Darwinian" writers in opposition to the lasting fame of unindebted originality that he attributes to Wordsworth. By reasserting the complaints earlier made against Darwin by certain critics, writers, and reviewers, Hazlitt makes his preferences clear and effectively consigns the movement of scientific literature to oblivion. In other words, by dividing poetry from the natural sciences and portraying Darwin as the now-recognizable straw man–representative of scientific literature, in order to elide the movement's actual diversity and complexity and pronounce it ephemeral and thus inconsequential, these now-canonical nineteenth-century male writers and literary critics ensured that scientific literature would be excluded from the canon.

In addition to Hazlitt's influence on the canonical erasures of Darwin and the movement of scientific literature, it is also worth noting his general dismissal of women writers in his evaluation of modern poets; he briefly and condescendingly discusses only three: Anna Barbauld, Hannah More, and Joanna Baillie. He calls Barbauld a "very pretty poetess" and associates her with "religious controversy" and children's literature from "when I was learning to spell words of one syllable"; he then wryly seems unsure of whether More is "still living" but assures that "she has written a great deal which I have never read"; finally, he infantilizes Baillie and her plays, explaining that "she treats her grown men and women as little girls treat their dolls."[12] Hazlitt's denigration of women writers is arguably typical of his critical perspective as he wrote, for instance, in his essay "Of Great and Little Things" (1821), "I have an utter aversion to *bluestockings*. I do not care a fig for any woman that knows even what *an author* means."[13] Thus, considering women writers' strong association with the movement of scientific literature, it is perhaps unsurprising that he chooses to exclude both women and the sciences from recognition within the developing standards of the canon. While Hazlitt also influenced De Quincey's efforts as a literary critic, the latter incorporated derogations of women and the sciences into his conceptions of the "literature of knowledge" and the "literature of power," formulations that he admits derive directly from Wordsworth.[14]

De Quincey first attempts to define literature by distinguishing power from knowledge in his "Letters to a Young Man Whose Education Has Been Neglected" (1823); this is a crucial text for understanding how now-canonical Romantic-era aesthetics sought to separate literature from science, championing the former and demeaning the latter. He ostensibly addresses "Letters" to a thirty-two-year-old man of wealth and status who now wishes to fill gaps in his education. Assuming authority within his role as literary critic, De Quincey emphasizes that he will offer guidance enabling the young man to decide on his own course of reading. In stating this necessity of choosing between disciplines, De Quincey foreshadows his division of literature from science and, interestingly, makes this separation integral to the preservation of sanity.

Explaining to the young man the need for selective reading, De Quincey relates that in his own youth he sometimes became so overwhelmed by the impossibility of obtaining complete knowledge that he suffered a kind of "madness." He writes, "In my youthful days, I never entered a great library, suppose of one hundred thousand volumes, but my predominant feeling was one of pain and disturbance of mind" because "here, said I, are one hundred thousand books—the worst of them capable of giving me some pleasure and instruction; and before I can have had time to extract the honey from one-twentieth of this hive, in all likelihood I shall be summoned away" (3:63–64). Realizing that his own mortality permits him to read only a fraction of the books in existence, he despairingly calculates that "ten thousand [books] for thirty years—will be as much as a man who lives for that only can hope to accomplish," so that in his entire life "the utmost he could hope to travel through would be twenty thousand volumes,—a number not, perhaps, above *five percent* of what the mere *current* literature of Europe would accumulate in that period of years" (3:64). Through this epiphany, De Quincey experiences what could be described as a distressing form of the sublime, feeling overwhelmed by the sheer volume of books and his own inability to read them all, thus producing "as real a case of suffering as ever can have existed" (3:64). His affliction then grew worse, "for the same panic seized upon me with respect to the works of art," including music and "other arts," as well as people themselves, as "it occurred to me that every man and woman was a most interesting book, if one knew how to read them," thereby opening "upon me a new world of misery; for, if books and works of art existed by millions, men existed by hundreds of millions" (3:64). De Quincey then recollects, "My madness took yet a higher flight; for I considered that I stood on a little isthmus of time, which connected the two great worlds, the past and the future," and "I asked for admittance to one as much as the other" so that "I fell into a downright midsummer madness. I could not enjoy what I had.—craving for that which I had not, and could not have" (3:65).

Analyzing his experience, De Quincey delineates his former "madness" in relation to other forms of insanity. He writes that his "panic" in realizing that he would never obtain full knowledge was "madness, I grant; but such a madness! not as lunatics suffer; no hallucination of the brain; but a madness" of his mind "travelling into excess" (3:65). De Quincey's description of his madness as an "excess" echoes, for example, Erasmus Darwin's definitions of insanity in his medical text *Zoonomia* (1794–96).[15] Yet, De Quincey asserts, "with allowance for difference of degrees, no madness is more common," and if some "do not carry it on to the same extremity of wretchedness, it is because they are not so logical, and so consistent in their madness, as I was" (3:65). Strikingly, he here counters the expectation of instability, insisting that his madness was "logical" and "consistent," and that if anyone does not experience this degree of "wretchedness," it is because they lack his understanding of the circumstances. De Quincey summarizes the problem, declaring that, "under our present enormous accumulation of books, I do affirm that a miserable distraction of choice (which is the germ of such madness) must be very generally incident to the times . . . and that one of the chief symptoms is an enormous 'gluttonism' for books, and for adding language to language; and in this way it is that literature becomes much more a source of torment than of pleasure" (3:65). However, according to him, while "choice" functions as "the germ of such madness," it also provides the solution.

To cure the "madness" resulting from this desire to read and know all things, De Quincey prescribes a deceptively simple remedy: "Dare to be ignorant" (3:69). He reveals that this phrase originated in Mathias's satirical poem *The Pursuits of Literature* (1794), which, as I discussed in the introduction, launched one of the first public attacks on Darwin's *Botanic Garden*, particularly condemning that scientific poem's "novelty" as well as its recommendation of botany as a study for women. De Quincey calls Mathias's exhortation "to *dare* to be ignorant of many things" a "wise counsel, and justly expressed; for it requires much courage to forsake popular paths of knowledge, merely upon a conviction that they are not favorable to the ultimate ends of knowledge" (3:69). Thus, he explains that the young man must embark on a "system" of reading that allows him to be "thoroughly possessed and occupied by the deep and genial pleasures of one truly intellectual pursuit" (3:69). Yet, in determining the young man's "one truly intellectual pursuit," a crucial choice must be made, for "you know whether your intentions lean most to science or literature" and "upon this decision" depends his future path (3:69). In this way, De Quincey makes it clear that one must *choose between* science and literature; one cannot and should not pursue both. The implications of this pronouncement regarding scientific literature are obvious, as such literature that seeks to combine these separate disciplines necessarily, in his formulation, becomes an overwhelming expression of "madness."

Clarifying the need to keep literature distinct from science, De Quincey defines his separate categories of "literature" and "knowledge." In fact, he writes that "literature is the direct and adequate antithesis of books of knowledge" (3:69). Discussing what he means by books of knowledge, De Quincey claims, "All knowledge may be commodiously distributed into science and erudition," including "books of voyages and travels" (3:72, 70). Yet he also asserts that "it is difficult to construct the idea of 'literature' with severe accuracy; for it is a fine art—the supreme fine art, and liable to the difficulties which attend such a subtle notion" (3:70). Thus, introducing the additional category of power, De Quincey suggests, "the true antithesis to knowledge" is "power," and "all that is literature seeks to communicate power; all that is not literature, to communicate knowledge" (3:70–71). Fascinatingly, he arguably aligns power with concepts of sensibility, defining the communication of power as that which makes one "feel vividly, and with a vital consciousness, emotions which ordinary life rarely or never supplies occasions for exciting, and which had previously lain unawakened" (3:71). Briefly pointing to Shakespeare's *King Lear* and Milton's *Paradise Lost* as examples of power, De Quincey makes no secret of his predilections; according to him, the communication of power is "far above all communication of knowledge," which he deems the "anti-literature" (3:71).

Crucially, De Quincey makes gender central both to his distinction between literature and science, and to the need to choose between these two disciplines and focus intently on "one truly intellectual pursuit." He aligns the natural sciences with women in an effort to make scientific fields appear emasculated, laughable, and disorganized, ridiculing the very fact of science's past inclusiveness or openness to women's participation. De Quincey thereby also subtly seeks to justify his more exclusive and masculinized conception of literature in association with "power." He explains,

> I remember once that, happening to spend an autumn in Ilfracombe, on the west coast of Devonshire, I found all the young ladies whom I knew busily employed on the study of marine botany. On the opposite shore of the channel, in all the South Welsh ports of Tenby, &c., they were no less busy upon conchology. In neither case from any previous love of the science, but simply availing themselves of their local advantages. Now, here a man must have been truly ill-natured to laugh; for the studies were in both instances beautiful. A love for it was created,

if it had not pre-existed; and, to women and young women, the very absence of all austere unity of purpose and self-determination was becoming and graceful. Yet, when this same levity and liability to casual impulses come forward in the acts and purposes of a man, I must own that I have often been unable to check myself in something like a contemptuous feeling. (3:62)

With a clearly sardonic and patronizing tone, De Quincey demeans women's studies of the natural sciences as ipso facto lacking "unity of purpose and self-determination," and as only guided by "levity" and "casual impulses," which would be "contemptuous" in a man. In doing so, he strives not only to emasculate the sciences but also to make the era's movement of scientific literature, in which so many women writers had participated, appear "contemptuous" and their texts unworthy of consideration or inclusion within a plan of reading for any young man and, thus, within any conception of the literary canon. Instead, he follows Wordsworth in masculinizing and appropriating emotions and sensibility, which had often been the terrain of women writers, within his formulation of the literature of power, and he separates "literature" from science so that the scientific literature of sensibility written by numerous men and women writers earlier in the era cannot fit within his categorizations; in fact, he specifically defines "literature" to exclude such writing.

De Quincey admits in a footnote of "Letters" the influence of Wordsworth on his thinking about the separation of knowledge from power. He writes, "For which distinction, as for most of the sound criticism on poetry, or any subject connected within it that I have ever met with, I must acknowledge my obligations to many years' conversation with Mr. Wordsworth" (3:70). Thus, through Wordsworth's sway over literary criticism in what became enduring texts defining both "Romantic aesthetics" and the literary canon, by critics such as De Quincey and Hazlitt, this older poet dictated the standards by which, not only his own writings, but all "literature" must be judged. Moreover, Wordsworth continued to assert his earlier views regarding the necessity of separating literature from science, as is evident, for instance, in his correspondence with William Rowan Hamilton, an aspiring poet who, in 1827, at twenty-two, was made the Andrews Professor of Astronomy at Trinity College, Dublin, and Astronomer Royal for Ireland. Hamilton met Wordsworth later that same year while visiting the Lake District, and the two men struck up a warm friendship. However, when the younger poet shared some of his verses, Wordsworth made it clear that Hamilton, who sought to combine literature and science in both his career path and his poetry, must choose between them. Differing "from his young friend in regarding these pursuits as competing rather than complementary activities," Wordsworth stressed that poetry and science each distinctly require a professional commitment and counseled Hamilton to maintain "the path of Science."[16] Thus, in the late 1820s, Wordsworth still was policing what he considered appropriate practices within poetry and excluded blendings that might resemble the movement of scientific literature.

While, in "Letters," De Quincey enshrines Wordsworth's principles to preclude science and women from the developing literary canon, his personal prejudices become even more apparent in his essay "False Distinctions" (1824). Here, De Quincey corrects the "errors" in what he calls three common "delusion[s]" regarding literature. Seeking to justify his essay's obvious discriminations based on sex and race, he writes that the three literary "delusion[s]" are: "That women have more imagination than men"; "that the savage has more imagination than the civilized man"; and "that Oriental nations have more imagination (and according to some a more passionate constitution of mind) than those of Europe."[17] The overt misogyny and racism of De Quincey's intentions within this essay alone should convince twenty-first-century scholars that he is an author whose critical perspectives should be either reviled or considered extremely suspect rather than repeatedly upheld as touchstones of Romantic-era aesthetics. Scholars such as Peter J. Kitson already have recognized De Quincey "as an 'ultra-racist' of the period in terms of the violence and extremism of his writing."[18] In "False Distinctions," De Quincey deprecates poetry authored by "savages" and suggests their supposed "intellectual barrenness" while more strongly condemning the literature from "Oriental nations" as lacking not only imagination but also sensibility or a "passionate constitution." Here, he particularly attacks "Arabian" literature, but later would similarly represent the Chinese as "less subject to pain because their nervous system was less sensitive than white Europeans," an assertion that "became a cliché of nineteenth-century racist discourse."[19] I will leave it to other scholars to further investigate the canonical consequences of the racism in De Quincey's literary criticism and, for the purposes of my current study, will more closely focus my examination on some of his vitriol against women writers. For example, in this essay, he demands:

> What work of imagination owing its birth to a woman can ["any man"] lay his hand on (I am a reasonable man, and do not ask for a hundred or a score, but will be content with one), which has exerted any memorable influence, such as history would notice, upon the mind of man? ... Where is the female rival of Chaucer, of Cervantes, of Calderon? Where is *Mrs.* Shakespeare?—No, no! good women: it is sufficient honor for you that you produce *us*—the men of this planet—who produce the books (the good ones, I mean). In some sense therefore you are grandmothers to all the intellectual excellence that does or will exist: and let *that* content you. As to poetry in its *highest* form, I never yet knew a woman—nor will believe that any has existed—who could rise to an entire sympathy with what is most excellent in that art. High abstractions, to which poetry ... is always tending, are utterly inapprehensible by the female mind; the concrete and the individual, fleshed in action and circumstance, are all that they can reach: ... the ideal—is above them.[20]

In these chauvinistic assertions, De Quincey ridicules the possibility of women ever achieving literary—and particularly poetic—"excellence." His arguments display sexist denial of the historical (and especially educational) obstacles endured by women in patriarchal societies, the sex-based critical expectations imposed on women writers, as well as women's numerous extant literary accomplishments. As Julian North observes of this passage, for De Quincey women's "primary creative function is biological, and subordinate to the life-giving male imagination."[21] Significantly, various eighteenth-century medical treatises interconnectedly associated women with sensibility and with the "Power of the Imagination" to affect even their offspring during the reproductive process, yet De Quincey attempts to separate women from their imaginative "power" of literature and sensibility and confine them to the domestic sphere and physiological reproduction.[22]

Denying women's excellence in the genre of poetry, De Quincey additionally ignores their great success as authors of novels. He perhaps alludes to novels as unworthy of mention when he states that men "produce the books (the good ones, I mean)," indicating that, in his view, novels represent an inferior form of writing.[23] As I have discussed elsewhere, Robert Southey earlier satirized novel-writing by depicting it in association with botany.[24] Moreover, in De Quincey's essay, although many women poets could be said to have explored what he calls "the ideal" or "high abstractions" in their work, he confines women's poetry to "the concrete and the individual, fleshed in action and circumstance"; his insistence on the inferiority of such poetry, again arguably targets women's scientific literature when relating the particulars and details of aspects of the natural world. However, it also should be noted that attention to the individual or particular does not preclude simultaneous or connected concern with the abstract, as is often evident in women's and men's scientific poetry. In fact, science itself depends on the ability to theorize generalities and abstractions from particulars. Yet De Quincey's denigration of, and efforts to erase, generations of women writers, especially in combination with similar statements from other contemporary male critics and reviewers, demonstrates how women writers (and, connectedly, scientific literature) were historically excluded from the developing standards of the literary canon.

In this way, De Quincey's assertions align with various remarks made by Wordsworth regarding women writers. For example, when William Rowan Hamilton's sister, Eliza, also displayed talent as a poet, Wordsworth declared that "female Authorship is to be shunned as bringing in its train more and heavier evils than have presented themselves to your sister's ingenuous mind."[25] His discouragement of women's writing is further clarified in a letter to his cousin, Elizabeth Fisher, regarding her daughter, Emmie:

> I have thought it proper both directly and indirectly to impress Emmie's mind with a conviction that talents and Genius, and intellectual acquirements, are of little worth, compared with the right management of the affections, and sound judgment in the conduct of life: that what she may become as a Woman, is of infinitely more importance than what she may grow into as a person of splendid intellect, or as an Authoress in any department of Literature. All this I have urged not merely for her own tranquility and happiness, but for much higher considerations of domestic and social duty, and religious obligation.[26]

While Deborah Kennedy describes Wordsworth as sometimes "cautiously supportive of aspiring female writers," those writers also recognized the limits of his support.[27] Felicia Hemans, for instance, diplomatically notes his "patriarchal simplicity," and Maria Jane Jewsbury more pointedly acknowledges that Wordsworth "seemed to think writing and femaleness a regrettable combination."[28] He extensively critiqued the manuscript poems of Catherine Grace Godwin so that she devoted considerable effort to "the task of rewriting her compositions," prompting her 1854 memoirist to lament that "some indeed may be of opinion that she guided herself too exclusively by Wordsworth's peculiar views of his art."²⁹ Yet, while demanding conformity to his poetic perspective, Wordsworth simultaneously complained of Hemans, for example, that she "felt herself under the necessity of expanding the thoughts of others, and hovering over their feelings, which has prevented her own genius doing justice to itself, and diminished the value of her productions accordingly."³⁰ He thus creates the necessity that younger women poets follow his "thoughts" and "feelings" and then condemns them for it. In this vein, Mary Russell Mitford wryly wrote in 1817 that she had been busy "trying to learn to admire Wordsworth's poetry[,] . . . to admire *en masse*—all, every page, every line, every word, every comma; to admire nothing else, and to admire all day long. This is what Mr. Wordsworth expects of his admirers (I had almost said his worshipers)."³¹

Wordsworth more scathingly critiqued earlier women authors. In 1808, Henry Crabb Robinson already observed Wordsworth's tendency to "quote . . . his own Verses with pleasure . . . [and] attach to the approbation of them a greater connection with moral worth which others may deem the effect of vanity" and to speak "with a contempt of others that I think very censurable. He asserts for instance that Mrs. B[arbauld] has a bad heart; that her writings are absolutely insignificant, her poems are mere trash and [that] specimens of every fault may be selected from them."³² Wordsworth made it clear that he also disapproved of Barbauld's beliefs and personal connections as a Dissenter. Although, as evidence of Wordsworth's interest in women writers, some scholars point to his offer in 1829 to contribute to Dionysius Lardner's *Cabinet Cyclopaedia* (1829–49) "an Account of the Deceased Poetesses of Great Britain—with an Estimate of their Works," he immediately withdrew this suggestion, claiming that "upon more mature Reflection" the subject is not "sufficiently interesting."³³ Nevertheless, he found it "very ungallant" that the editor of no "other Corpus of English Poetry takes the least notice of female Writers" and congratulated the Scot editor and literary historian Rev. Alexander Dyce on his 1825 one-volume *Specimens of British Poetesses*. Corresponding with Dyce later in 1829, Wordsworth writes that if a second edition should be required then he would like to be consulted about it, particularly regarding Anne Finch, Helen Maria Williams, Charlotte Smith, Anna Seward, and Anna Barbauld, despite admitting that he only possesses "scanty materials" within his personal library to aid him in this task.³⁴ As Stephen Behrendt notes, Wordsworth's "slim library of women poets" displays that he did not actually collect or seek out women's works, and that "it is clear that poetry by his female contemporaries was neither a priority nor even a particularly strong interest with Wordsworth."[35] This male poet also undermines the significance and success of these women writers, stating, for instance, that Seward's "verses please me, with all their faults, better than those of Mrs. Barbauld, who, with much higher powers of mind, was spoiled as a poetess by being a dissenter," and places Finch, whom he perhaps praises the most highly of these women, only on a par with the "minor poet" Thomas Tickell.[36]

In Wordsworth's letters to Dyce about women writers, as in his earlier preface to *Lyrical Ballads*, he draws botanical comparisons with literature, even as he ostensibly rejects scientific literature itself. For example, he associates Finch's poems with the uncultivated flowers of the fields, as opposed to florists' artificial, "monstrous," "double flowers": "Her style in rhyme is often admirable, chaste, tender, and vigorous; entirely free from sparkle, antithesis, and that over-culture which reminds one—by its broad glare, its stiffness, and heaviness—of the double daisies of the garden, compared with their modest and sensitive kindred in the fields."[37] Dyce may have prompted these botanical analogies from Wordsworth by asking whether his poem "To Enterprise" (1820) drew on Erasmus Darwin's verses. Wordsworth confirmed this fact, admitting that the "obligation to Dr. Darwin.... was acknowledged in a note, which slipped out of its place" in the poem's most recent edition.[38] This displays both Darwin's continued influence on Wordsworth, as well as the younger poet's conflicted disavowal of his indebtedness to Darwin and of his own works' intertextuality with scientific thought. Indeed, Wordsworth suggests to Dyce poems by several contemporary women writers (Smith, Seward, Williams, Barbauld) who variously engaged with technical science, and especially natural history; yet, he recommends verses from these women's oeuvres that arguably display their sensibility and not their facility with the natural sciences, imposing a posthumous separation of these often conjoined aspects of their verse. In other words, if these now-deceased women writers were to be remembered, then he makes a point of exerting control over how their reputations should be constructed and which of their works might be preserved. Since "Wordsworth was firmly established in 1829 as a—perhaps *the*—preeminent living British poet," he could feel confident by this time in the success of his (masculinized) methods of poetry and efforts (in theory, if not always in practice) to separate science from verse; he could now afford to be more generous toward the memories of these women writers and their works, especially when he could selectively shape them to conform to his poetic vision.[39] Disheartingly, in addition to Wordsworth's influence on De Quincey's critical views of women, other male writers, critics, and reviewers during the period also endorsed such assertions. John Clare, for instance, readily agreed that De Quincey's essay "False Distinctions" had "contended right enough that women had an inferior genius to men."[40] That essay also anticipates similar statements from other male authors and critics during the Victorian era, including Ruskin's famous lecture "Of Queens' Gardens" (1864), in which he suggests that if women would seek "to heal, to redeem, to guide and to guard," rather than to obtain the economic and educational advantages of men, then they would become "no more housewives, but queens."[41] Both De Quincey's "Letters" and "False Distinctions" provide important insights for understanding the motivations and prejudices underlying his more often anthologized 1848 discussion of knowledge and power.

In his review essay "The Works of Alexander Pope" (1848), De Quincey revises his division between the "literature of knowledge" and the "literature of power" to place greater emphasis on the importance of originality and sensibility. While some scholars view this later essay as "soften[ing]" his distinction between knowledge and power, noticeable problems persist in his definitions of these categories.[42] For example, continuing to align power with sensibility, De Quincey claims that the function of the literature of knowledge is "to *teach*," while that of the literature of power is "to *move*," clarifying that it "teaches, if at all, indirectly *by* moving."[43] He explains that with knowledge (which still prominently includes the sciences) the reader remains earth-bound, as it will only "carry you further on the same plane," while with power the reader ascends to "mysterious altitudes above the earth," as this literature expands "human sensibilities" and "moral capacities" by connecting readers with the "infinite" through the "great intuitive" organ of the "heart" (16:337). However, as I have shown throughout *Regenerating Romanticism*, writers of scientific literature often expressed these imaginative, emotional, and moral facets of sensibility that De Quincey associates with power; thus, his distinction merely repeats the earlier attacks on Darwin's poetry as lacking sensibility, and that then falsely imputed a dearth of feeling to scientific literature more generally, as discussed in the introduction. Moreover, scholars such as Noel Jackson assert the primacy of teaching in Wordsworth's own verse, stating that "Wordsworth makes bodily feeling indissociable from, and indeed often 'subservient' to, the didactic purposes of the poet's craft."[44] In fact, in Wordsworth's 1815

"Essay, Supplementary to the Preface," he declares that, for an original author, "to create taste is to call forth and bestow power, of which knowledge is the effect" (660). Since De Quincey credits Wordsworth with coining these concepts and with exemplifying power, the older poet's statement here suggests that in power there is an overt intention to teach, to bestow "knowledge." Thus, as he did in "Letters," De Quincey is already exhibiting "difficulties" in distinguishing between knowledge and power.

Like Wordsworth in his 1802 preface, De Quincey employs plant imagery within his 1848 essay; yet, while Wordsworth sometimes paradoxically (or contradictorily) appropriates botanical references to reinforce his literary assertions despite ostensibly rejecting the intermixing of science and literature, De Quincey pointedly delimits botany's relation to his ideal notion of literature. Reprising others' accusations from half a century earlier that Darwin's poetry conveys "mere novelty," De Quincey connects "seeking information or gaining knowledge" with "absolute novelty," but writes that the "higher" truth of power "is never absolutely novel"; instead, power "exists eternally by way of germ or latent principle...needing to be developed, but never to be planted. To be capable of transplantation is the immediate criterion of a truth that ranges on a lower scale," and power is "deep sympathy with truth" (16:336). In other words, according to De Quincey, power awakens emotions and thoughts already present within the reader, while writers of knowledge must act as botanists or gardeners and "plant" within the reader the desired information or ideas. However, as I showed in *Questioning Nature* and discussed in the introduction to this book, many writers of scientific literature were, in fact, very concerned with achieving originality as well as novelty in their literary works, as were Wordsworth and Coleridge, making De Quincey's distinction based on novelty another false divide. Continuing to demean knowledge in connection with "novelty," he exemplifies power through Milton, asking, "What do you learn from Paradise Lost? Nothing at all. What do you learn from a cookery book? Something new, something that you did not know before, in every paragraph. But would you therefore put the wretched cookery book on a higher level of estimation than the divine poem?" (16:337). De Quincey's dismissal of knowledge and novelty through a "cookery book" arguably has gendered implications as many cookbooks of the era were written by and/or for women.[45] Moreover, his association of gardening and cooking with the literature of knowledge subtly seeks to reduce (especially women's) botanical literature to the mere use of vegetables in domestic, practical terms.

Significantly, in this vein, De Quincey's invectives against women did not alter in the decades following his essay "False Distinctions." In fact, during the year prior to this 1848 essay on knowledge and power, he published his essay "Joan of Arc" (1847), where he similarly asserts, "Woman, sister—there are some things which you do not execute as well as your brother, man; no, nor ever will. Pardon me if I doubt whether you will ever produce a great poet from your choirs, or a Mozart, or a Phidias, or a Michael Angelo, or a great philosopher, or a great scholar" (16:82). De Quincey's continued denial of women's intellectual capacities here suggests that women's writing (and especially their scientific literature, through association with knowledge) remained a primary target of attack in his 1848 essay, dividing knowledge from power. Indeed, modern scholars have recognized "deep conflicts" in his writings about women and "anxious assertions of masculine power."[46] John Barrell notes in him a "contempt for the effeminate" that seems to betray "a disavowal of identification with the feminine."[47] For Tim Fulford, De Quincey's writing shows a "fear of the feminine" that provokes assertions of masculine agency and "power," and John Whale similarly sees the theory of power as a compensation for "the inherent femininity" of "the realm of letters," evincing "a deep-seated anxiety about the degree to which the profession of literature might constitute a threat to one's manliness."[48]

In his 1848 essay, De Quincey further undermines his arguments by creating false or erroneous alignments of power with the aesthetic fantasy of unindebted literary originality, as opposed to knowledge's association with intertextuality. He posits of knowledge, "Let its teaching be even partially revised, let it be but expanded—nay, even let its teaching be but placed in a better order—and instantly it is superseded" (16:338). He exemplifies Isaac Newton's *Principia* "as a living *power* [that] has transmigrated into other forms" through the work of La Place and others, throwing Newton's work into "decay and darkness" so that "the name of Newton remains as a mere *nominis umbra*," or shadow of a name (16:338). Here, not only does De Quincey inconsistently attribute "power" to Newton's text of scientific "knowledge," but he also asserts a "false distinction" (to appropriate his earlier essay's title) by demeaning the mutability of forms and ideas in the literature of knowledge. Although eighteenth- and early nineteenth-century scientific texts were often revised, expanded, or "placed in a better order" through new editions, the same, of course, could be said of the works of now-canonical male Romantic-era poets who also often revised, expanded, or reordered their works. Additionally, while it may be true that historians now comprise the main readers of Newton's *Principia*, his name arguably has proven as "immortal" as Milton's or Shakespeare's through the continued importance of his discoveries. On the other hand, De Quincey claims that literature of power, such as *Hamlet*, *Macbeth*, and *Paradise Lost*, is "finished and unalterable" and "never *can* transmigrate into new incarnations" because "to reproduce *these* in new forms, or variations, even if in some things they should be improved, would be to plagiarize" (16:338). Since all of the "big six" male Romantic poets borrowed, "improved," and practiced additional forms of intertextuality, sometimes drawing specifically on Shakespeare or Milton (who each themselves also adapted or transformed earlier works), it is difficult to feel convinced by De Quincey's differentiations of the literature of knowledge from that of power; however, it is easy to see how his assertions reinforced misrepresentations of those now-canonical male Romantic-era poets as fulfilling the myths of solitary genius and autonomous or unindebted originality.

Again, in making these distinctions that claim the superiority of power through its alignment with originality, De Quincey's primary (and, in many ways, conjoined) targets are scientific literature and women writers—and, paradoxically, his arguments about originality draw on and mirror some related scientific assertions about sex and race. Throughout the nineteenth century, both male literary critics and scientific thinkers frequently supported "long-standing cultural perceptions of an innate female inferiority," and, within this supposed sexual distinction, men were portrayed as excelling in complex abstractions, reason, judgment, imagination, and originality while women were restricted to intuition, perception, and imitation.[49] In this regard, the era's scientists, such as Charles Darwin, in his *Descent of Man, and Selection in Relation to Sex* (1871), deemed those traits of imitation, and so on, as not only associated with women but also as "characteristic of the lower races, and therefore of a past and lower state of civilization."[50] Such "scientific" statements further display ways in which sex and race became connected through masculine Eurocentric efforts to control and confine participation in conceptions of originality and sensibility. Apparently unconcerned by his own lack of originality, De Quincey copies assertions by both Wordsworth and Hazlitt, stating that the ultimate difference between literature of knowledge and that of power is time, suggesting that the former "passeth away" while the latter is "much more durable" (16:339). Of course, his statement also strategically elides the role of literary critics such as himself in determining which texts endure and which "passeth away."

Nevertheless, although De Quincey still asserts the opposition or "reciprocal repulsion" of the literatures of knowledge and of power, unlike in "Letters" he here briefly acknowledges that these categories "may blend and often *do* so" (16:336). He only explores this blending in a footnote, explaining, "a vast proportion of books—history, biography, travels, miscellaneous essays, etc.—lying in the middle zone, confound these distinctions by interblending them"; such books constitute what "we call 'amusement' or 'entertainment' [which] is a diluted form of the power belonging to passion, and also a mixed form. . . . or neutral state [in which knowledge and power] disappear to the popular eye as the repelling forces which, in fact, they are" (16:339). Perhaps the scientific literature on which I have focused my study, and which often conveys knowledge as well as the originality and sensibility that De Quincey associates with power, fits within this gap in his attempt to formulate a more polarizing literary taxonomy between knowledge and power; however, if so, then it is equally important to recognize that the same could be said of the literature of the now-canonical male Romantic poets. As I have shown through the inconsistencies in both Wordsworth's and De Quincey's literary theories, the literature of power constitutes a constructed category, like that of solitary genius or the autonomous originality on which this category depends, and is thus yet another aesthetic fantasy of the era. Instead, the now-canonical male Romantic poets, like many of the imaginative authors who engaged more obviously with the natural sciences and whom I particularly analyze in this book, wrote variously within the spectrum of this blending, arguably collapsing or rendering futile De Quincey's divisions between knowledge and power. As opposed to De Quincey's false conception of power, for the women and men writing scientific literature, knowledge could afford cultural, sociopolitical, economic, and authorial power or authority, especially through the combination of science and sensibility; and, as I have discussed, this was a power that now-canonical male Romantic poets recognized and therefore often selectively repudiated and appropriated for their own purposes.

Thus, despite the supposed "soften[ing]" of De Quincey's distinctions between the literature of knowledge and that of power in his 1848 essay, he maintains this (erroneous) theoretical division; and, again, it is imperative to understand his essay in the context of its earlier manifestations in "Letters" and "False Distinctions," which more strongly display his prejudicial motivations and desire to *disempower* and exclude women and non-European authors, as well as (and, in the cases of many women, as representatives of) writers of scientific literature. In doing so, De Quincey supports Wordsworth's attempts to appropriate and monopolize sensibility and originality through his category of the literature of power. As I have shown, what became canonical Romantic-era aesthetics in many ways developed specifically in opposition to the movement of scientific literature, even as each of the now-canonical male Romantic poets also engaged with the natural sciences in various and individual ways. Further, this is not to say that the late eighteenth- and early nineteenth-century movement of scientific literature itself did not sometimes also interact with, or even incorporate, sexist or racist thought. As displayed, for instance, in chapters 1 and 3, forms of scientific racism were sometimes exhibited through ideas about stadial theory, interconnections of biological organisms with climate and the environment, and numerous additional "scientific" conventions of the era; while, at the same time, such contexts also enabled some writers to instead envision, for example, possibilities for universal human equality. Moreover, not only did Linnaeus's eighteenth-century botanical taxonomies include biases based on gender and sex, as I have discussed, but also, as other scholars have demonstrated, his natural history texts categorized humanity into four racial groups: "American, European, Asiatic, and African . . . modeled on the four elements (earth, air, fire, and water) and the four corners of the world created by the four rivers of Eden"—classifications that were "criticized at the time for their arbitrariness and essentialism," yet "began a trend of racial thinking" regarding "fixed racial types."[51] Nevertheless, as I have explored in this book, male and female writers within the movement of scientific literature often questioned, corrected, negotiated, and revised in various ways the assertions of contemporary naturalists, providing the possibility of reconceptualizing the era's scientific constructions of identity categories including gender, sex, and race.[52]

Within this broader cultural perspective, it is therefore tempting to wonder what might have occurred if science and literature had continued to move toward embracing the often more inclusive and openly collaborative values of (and between) these disciplines that had been present in the decades prior to professionalization and that had made the movement of scientific literature possible. Had these fields maintained such possibilities for inclusivity and mutual accountability, rather than frequently mirroring one another's prejudices and enacting disciplinary competition as the nineteenth century progressed, the course of history and treatments of gender, sex, and race in these professions may have evolved differently and perhaps in a more genuinely compassionate manner. However, while we cannot change the past, we can strive to see it more clearly and then alter or remediate our own scholarly and pedagogical practices accordingly. As Devoney Looser points out regarding late eighteenth-century and Romantic-era women authors, "A generation or two later, these once-dominant writers had become difficult to trace" so that, "by the mid-nineteenth century, Elizabeth Barrett Browning could not locate any literary 'grandmothers' among British poets."[53] Similar statements register the historical erasures of this era's writers of scientific literature.[54] As I have shown, in both (interrelated) cases, such elisions resulted in many ways from converging and often retroactive and erroneous efforts of nineteenth-century literary critics, reviewers, now-canonical male authors, and scientific theorists seeking to control claims to authority through concepts such as knowledge and power, as well as sensibility and originality.

Regenerating Romanticism strives to provide a way forward in the scholarship and teaching of late eighteenth- and early nineteenth-century British literature, a significant era of canon formation that arguably set the principles that would influence subsequent generations in the creation and evaluation of literature. More fully acknowledging how and why the exclusions and misrepresentations of historically marginalized authors (including women, non-Europeans, writers of scientific literature, and so on) occurred and were often retroactively imposed allows us to more accurately assess and recount the actions and motives of those involved. Genuinely appreciating, critically and fairly, each author on his or her own terms, while also holding those writers individually accountable, will enable us to achieve a more equitable and comprehensive understanding of this diverse and exciting era of study.

NOTES

Introduction

1. Coleridge to Southey, 29 July 1802, in Coleridge, *Collected Letters*, 2:830.
2. *British Critic* 14 (1799): 365; quoted in King-Hele, *Erasmus Darwin and the Romantic Poets*, 71.
3. *Edinburgh Magazine* 2 (1818): 313; quoted in King-Hele, 71. See also Moir, *Sketches of the Poetical Literature*, 12.
4. Thomas De Quincey, "Letters to a Young Man," in *Works*, 3:71. All references to this work will be cited parenthetically.
5. Wordsworth, *Major Works*, 607. All references to this work will be cited parenthetically.
6. Ross, *Contours of Masculine Desire*; Mellor, *Romanticism and Gender*.
7. McGann, *Romantic Ideology*.
8. British Romanticism's "big six" male writers typically consist of William Wordsworth, S. T. Coleridge, William Blake, Percy Shelley, George Gordon Byron, and John Keats.
9. On scientific literature's originality, see Bailes, *Questioning Nature*; on Erasmus Darwin's denigrated "novelty," see Bewell, *Natures in Translation*, 53–86; on scientific literature's supposed lack of sensibility, see Jackson, "Rhyme and Reason"; King-Hele also alludes to some of these aspects of Romantic aesthetics in *Erasmus Darwin and the Romantic Poets*.
10. Bewell, "Erasmus Darwin's Cosmopolitan Nature," 19.
11. See, for instance, the three texts cited above by Bewell, Jackson, and King-Hele, as well as Packham's "Science and Poetry of Animation." In "Twin Labourers and Heirs," Ross suggests that Wordsworth's rivalry with Humphry Davy additionally may have influenced his remarks. See also Ruston, *Creating Romanticism*, 20–27.
12. Coleridge, *Biographia Literaria*, 207.
13. See, for example, Thomson's *The Seasons* (1726–46); Whitehead's *The Goat's Beard: A Fable* (1777); Scott's *Amoebaean Eclogues* (1782); Seward's *Elegy on Captain Cook* (1780); Smith's *Elegiac Sonnets* (1786); Barbauld's "To Mrs. P—, with Drawings of Birds and Insects" (1773); Crabbe's *The Library* (1781); Cowper's *The Task* (1785); and Williams's *Peru* (1784).
14. It is worth noting that Ross advocated this approach particularly in regard to women writers in *Contours of Masculine Desire*.

15. For this and preceding sentences in this paragraph, see Bailes, *Questioning Nature*, 4, 7.
16. Aikin, *Essay*, 10.
17. Aikin, 25.
18. Bailes, *Questioning Nature*, 10. For various theories about the causes of the disappearance of the georgic, see Feingold, *Nature and Society*, 1; Crawford, *Poetry, Enclosure, and the Vernacular Landscape*, 92; Siskin, *Historicity of Romantic Discourse*, 12; and Goodman, *Georgic Modernity and British Romanticism*, 10.
19. Aikin, *Essay on the Application*, 41, 132. In *Science, Form, and the Problem of Induction*, Porter writes that Darwin's *Loves of the Plants* "couples the georgic's stated didactic end with the heroic couplets and speculative notes of the philosophical poem" (78); as she references, of course some georgics contained scientific or philosophical notes as well.
20. For this and preceding sentences in this paragraph, see Bailes, *Questioning Nature*, 11, 2, 4.
21. For the ideas in this paragraph, see, especially, Bailes, *Questioning Nature*, 2, 4, 14, 179–80; regarding the critical shift toward championing autonomous authorship, see also Macfarlane, *Original Copy*.
22. Aikin, *Essay* 34.
23. Aikin, 59.
24. Aikin, 132.
25. Aikin, 122–23.
26. Aikin, *General Biography*, 10:305; emphasis mine.
27. Aikin, 10:304; emphasis mine.
28. Seward, *Letters*, 6:157.
29. See also, for instance, Bailes, *Questioning Nature*, 114.
30. Aikin, *General Biography*, 304.
31. Ahern, *Affected Sensibilities*, 12; Ellis, *Politics of Sensibility*, 221; McGann, *Poetics of Sensibility*, 96; Pinch, *Strange Fits of Passion*, 11; Yahav, *Feeling Time*, 11.
32. Novak and Mellor, "Introduction,", 11.
33. Ahern, *Affected Sensibilities*, 11.
34. Keith, "Poetry, Sentiment, and Sensibility," 127. This paragraph gains direction from Keith's essay.
35. Mullan, *Sentiment and Sociability*, 236; Mullan, "Hypochondria and Hysteria," 141. See also Csengei, *Sympathy, Sensibility*. For a discussion of the "unavoidably paradoxical nature" of "emotion," both during the Romantic era and more generally, see Faflak and Sha, "Introduction."
36. Keith, "Poetry, Sentiment, and Sensibility," 128.
37. Barker-Benfield, *Culture of Sensibility*, xvii–viii; Johnson, *Equivocal Beings*, 14.
38. Novak and Mellor, "Introduction," 15, 16.
39. Novak and Mellor, 17.

40. Van Sant, *Eighteenth-Century Sensibility*, 4.
41. Rousseau, "Nerves, Spirits, and Fibres," 141n8.
42. Barker-Benfield, *Culture of Sensibility*, xvii; Ellis, *Politics of Sensibility*, 7.
43. Ahern, *Affected Sensibilities*, 21, 26.
44. Richardson, "Romanticism and the Colonization of the Feminine," 13.
45. Curran, "I Altered," 195, 197.
46. Chandler, "Question of Sensibility," 489. Chandler argues that "Romanticism was grounded in a notion of sensibility understood as a matter of generalized cultural crisis," of *in*sensibility due to public desensitization at a time of war, "urbanization, the monotony of the modern workplace, political upheavals, technological speedup," and so on (489, 475). He suggests that concepts of sensibility shifted during the Romantic period, and that "Wordsworth relocated the question of sensibility from the domain of metaphysics or theology (as in More, Sterne, and the Latitudinarians), or from a generalized theory of moral sentiments to a new frame of reference: a specific social and cultural situation grounded in the conditions of a historical moment" (476). However, I would contend that the location of sensibility in "a specific social and cultural situation grounded in the conditions of a historical moment" is not "new" with Wordsworth and in fact describes much of the poetry of sensibility in the last decades of the eighteenth century and throughout the Romantic era, especially by women authors, who often wrote in reaction to, or commemoration of, "specific social and cultural situation[s]," frequently in combination with moral concerns, and of course Wordsworth emphasized morality as well. Thus, as other scholars have suggested, Wordsworth appropriates sensibility for masculine poetics, while ignoring the poets, many of them women, who were already writing within that mode.
47. Curran, "I Altered," 198, 202.
48. See Wordsworth, *Poetical Works*, 4:403.
49. Curran, "I Altered," 202; Labbe, *Writing Romanticism*, 5.
50. Behrendt, "William Wordsworth and Women Poets," 640. Regarding the flourishing of this kind of poetry beginning around the 1740s, see, for example, Sitter, *Literary Loneliness*.
51. Mellor, *Romanticism and Gender*, 23, 29.
52. Mellor, 27.
53. Wordsworth, "Notes," in *Lyrical Ballads*, n.p.; Jackson, *Science and Sensation*, 2.
54. Jackson, "Rhyme and Reason," 172; Reiman, "Introduction," xiv; Bewell, "Erasmus Darwin's Cosmopolitan Nature," 20. See also Mitchell, "Cryptogamia," 648n15. Fara similarly observes that the male Romantic poets viewed Darwin as "their chosen representative of an outdated poetic movement" against whom they could "forge their own identity" (*Erasmus Darwin*, 44).

55. Darwin, *Life of Erasmus Darwin*, 33, 34.
56. Darwin, 34. Powell also discusses Erasmus Darwin's emphasis on the eye in "Linnaeus, Analogy, and Taxonomy."
57. Darwin, 31.
58. Darwin, *Botanic Garden, Part II*, vi.
59. Darwin, 43.
60. Darwin, 84.
61. Darwin, 49.
62. Coleridge, *Letters*, 1:177.
63. Seward, *Memoirs*, 178, 355; *Analytical Review* 15 (1793): 287–93. In *Nature's Body*, Schiebinger remarks that "though Anna Seward, a close friend and well-known poet, praised *The Botanic Garden* for establishing a new poetic form by adapting scientific discoveries to heroic verse, Darwin's poetry was not new, nor was it esoteric or unusual. The eighteenth century abounded with didactic poems on raising hops, sugar cane, gardening and the like" (31). Nevertheless, it seems scholars have yet to find an earlier poem that sought so thoroughly to versify the Linnaean botanical system, so there does seem to be a claim for Darwin's originality in that regard. In *Poetry of Erasmus Darwin*, Priestman writes that "Darwin's amorous plants had poetic precedents in the post-Augustan Roman poet Claudian . . . and in two English neo-Latin poems: Abraham Cowley's *Plantarum* (1662) and Demetrius de la Croix's *Connubia Florum* (1723). However, none of these poetic ancestors' toyings with plant sex had anything like the real-world impact of Linnaeus" (71).
64. See King-Hele, *Erasmus Darwin: A Life*, 238.
65. Stock, *Memoirs*, 131, 133.
66. Wordsworth, *Poetical Works*, 3:442.
67. Mathias, *Pursuits of Literature*, 1:5, lines 109–12. See also List, "Sometimes a Stamen," 205.
68. Mathias, 1:14.
69. Darwin, *Temple of Nature*, 83.
70. Darwin, 86.
71. Coleridge, *Biographia Literaria*, 207.
72. Coleridge, 208.
73. Wordsworth, *Lyrical Ballads*, 2:106.
74. Coleridge, *Biographia Literaria*, 13.
75. Coleridge, *Lectures*, 1:217.
76. Wordsworth, *Lyrical Ballads*, 2:166, lines 17–20.
77. Trott, "Wordsworth's Loves of the Plants," 148.
78. Wordsworth, *Poems*, 62.
79. *Anti-Jacobin Review and Magazine*, 5:246–47.
80. *Anti-Jacobin Review and Magazine*, 5:253, 255.

81. *Anti-Jacobin Review and Magazine*, 5:252, 255.
82. Jackson, "Rhyme and Reason," 193.
83. Keith, "Poetry, Sentiment, and Sensibility," 131.
84. Rothstein, *Restoration and Eighteenth-Century Poetry*, 100.
85. Keith, "Poetry, Sentiment, and Sensibility," 132.
86. Packham, "Science and Poetry of Animation," 195.
87. Packham, 197.
88. Seward, *Letters*, 5:378; Labbe, *Charlotte Smith*, 8; see also Bailes, *Questioning Nature*, 96.
89. Seward, *Memoirs*, 128, 110.
90. Coleridge, *Biographia Literaria*, 209.
91. Coleridge, 211. With this greater emphasis on pleasure rather than truth, Coleridge arguably seeks to differentiate poetry from its conception within scientific literature, as John Aikin, for instance, asserted that truth is crucial to poetic beauty (*Essay on the Application*, 25).
92. Drouin and Bensaude-Vincent, "Nature for the People," 408. See also Bailes, *Questioning Nature*, 121, 109.
93. See Ross, *Contours of Masculine Desire*, 50–51; emphasis mine.
94. Rousseau, "Nerves, Spirits and Fibres," 142n10.
95. Ahern, *Affected Sensibilities*, 16. See also Richardson, *British Romanticism*, 8; and Jordanova, *Sexual Visions*, 27–28.
96. Riskin, *Science in the Age of Sensibility*, 283, 287.
97. Each of these scholars has published multiple important texts and articles on botany, but to offer a few representative works, see Shteir, *Cultivating Women, Cultivating Science*; Schiebinger, *Nature's Body*; Bewell, "Jacobin Plants"; George, *Botany*; and Kelley, *Clandestine Marriage*. See also Page and Smith, *Women, Literature, and the Domesticated Landscape*; King, *Bloom*; and Sagal, *Botanical Entanglements*.
98. Kelley, 19.
99. Mitchell discusses those nonflowering plants to theorize his conception of male Romantic poets' approaches to plants more generally in "Cryptogamia," 631–51.
100. Schiebinge *Nature's Body*, 17.
101. Shteir, *Cultivating Women*, 16.
102. Schiebinger, "Private Life of Plants," 130.
103. Shteir, *Cultivating Women*, 18.
104. Shteir, 21.
105. Shteir, 24, 37.
106. Coleridge, *Notebooks*, 2:2601. For some verses perhaps written by Coleridge that sneer at women who study botany, see Trott, "Wordsworth's Loves of the Plants," 149. See also Fulford, "Coleridge, Darwin, Linnaeus," 127.
107. See Shteir, *Cultivating Women*, 83, 85.

108. Lucy Aikin, *Memoir*, 1:37.
109. Lucy Aikin, 1:38.
110. Darwin, *Loves of the Plants*, 149.
111. As Porter explains in *Science, Form, and the Problem of Induction*, Wordsworth defines the phrase "composite orders" in the preface to his *Poetical Works* (1827) as a combination of descriptive, didactic, and philosophical-satirical poetic modes and exemplifies Edward Young's *Night Thoughts* (1742) and William Cowper's *The Task* (1785). Although, "in 1815, Wordsworth had used the phrase 'composite species,' suggesting the botanical and typological roots of his thinking on poetic kinds," in 1827 he changed the phrase to "composite order" and thus "shifted from a taxonomic to an architectural metaphor" (60).
112. For especially striking examples of this kind of visible hybridity within zoology, one could think of a zonkey or liger, and of course there are many such hybrids within the plant kingdom.

1. Botany's Seasonal Disorder

1. Shteir, *Cultivating Women*, 240n4; Haut, "Reading Flora," 246n22.
2. Crawford, *Poetry, Enclosure, and the Vernacular Landscape*, 92. As Goodman writes, "We forget its phenomenal popularity both during the decades the several *Seasons* were first published and for almost a century afterwards" (*Georgic Modernity and British Romanticism*, 42).
3. On the literature of sensibility as reacting against Hobbes and Mandeville, see Novak and Mellor, *Passionate Encounters*, 12; and Ellis, *Politics of Sensibility*, 11. On the complexity of Mandeville's ideological relationship with sensibility, and particularly with sympathy, see Lamb, *Evolution of Sympathy*.
4. Thomson, *Seasons and Castle of Indolence*, Summer, l. 878. All references to *The Seasons* will be from this edition and cited parenthetically by poem and line number in the text. It's interesting to note that this temporal paradox became pervasive in the era for, as Gould argues, for instance, the discovery of geological deep time, between the seventeenth and nineteenth centuries depended on the mixing of two metaphors: time as a *cycle* of recurrence and time as an *arrow* of directional sequence and unrepeatable events (*Time's Arrow, Time's Cycle*). See also Menely, "Ecologies of Time," especially 86–87. This strikingly plays out in the geological debate between uniformitarianism and catastrophism. Likewise in British politics, the tradition of pro-Stuart and Jacobite typology emphasized history's cyclical nature, anticipating the dynasty's return, while Whigs such as Thomson tended to view history and temporality in a more directional and progressive manner; see, for example, Pittock, *Poetry and Jacobite Politics*, 10.

5. See, for example, Nicholson, "Fugitive Pieces," 142. See also Chakrabarty's conception of "historical time" in *Provincializing Europe*, 74.
6. Although some critics have treated a couple of these Backwardness of Spring poems individually, I do not know of any scholarship that has explored them together as a broader subgenre.
7. Chico, *Experimental Imagination*, 155–56. For more on Thomson's Newtonianism and Whiggish politics, see, for example, Willan, "Proper Study of Mankind"; Connell, "Newtonian Physico-Theology"; and Holberton, "James Thomson's *The Seasons*."
8. Aikin, *Essay on the Plan and Character*, iii, iv. Cohen notes that a "major effort to convert Thomson's poem to unity of a new kind was made by John Aikin . . . based on the assumption that the poem began a new genre, descriptive poetry" (*Art of Discrimination*, 88).
9. Aikin, *Essay on the Application*, 57.
10. Pennant, *Arctic Zoology*, vol. 1, cv.
11. Bailes, *Questioning Nature*, 10. Mahood also suggests this in *Poet as Botanist*, 20.
12. Bergstrom, "James Thomson's *A Poem*," especially 41–42.
13. Golinski, *British Weather*, 42.
14. Kaul, *Poems of Nation*, 149, 134.
15. Addison, *Spectator* no. 69.
16. Sambrook, *James Thomson*, 53.
17. In fact, Thomson directly associates Britain's beneficial climate and time when he refers to the nation's "temperate hours" (*Spring*, l. 443). In an interesting blend of climate and temporality, he references "Hours" which, in Greek mythology, were the goddesses of the weather who supposedly presided over changes in the seasons (*Summer*, l. 4).
18. Aikin, *Calendar of Nature*, 25.
19. Similarly, he then records the temporal order in which you might hear particular birds' songs during the day. Thus, "up springs the lark, / Shrill-voiced and loud, the messenger of morn," followed by the thrush, woodlark, blackbird, bullfinch, linnets, the jay, etc. (*Spring*, ll. 590–91, 598–610).
20. McKillop, *Background of Thomson's Seasons*, 55.
21. McKillop, 55.
22. McKillop, 55.
23. Addison, *Spectator*, no. 387; see also McKillop, 57.
24. As McKillop writes, "Thomson appears to be on both sides of the great question of primitivism versus progress. He is a Whig enthusiast for progress and a sentimental enthusiast for simplicity and primitive virtue. In one form or another this inconsistency was as old as Lucretius," and prevalent throughout the eighteenth century (89). Spacks writes that "the inconsistencies in Thomson's attitude toward progress and civilization . . . correspond

exactly to the inconsistencies in his attitude toward the state of natural man ... they suggest a complete ambiguity in his position about the relative virtue of natural and civilized man" (*Varied God*, 174–75); however, Thomson's prejudices are arguably more clear than this suggests.
25. Taylor, *Sky of Our Manufacture*, 10.
26. See Tobin, *Colonizing Nature*, 32. Later repeating this trope, Pennant, for example, writes: "The Torrid Zone generally enervates the body and mind. The divine particle melts away, and every idea is too often lost in irresistible indolence" (*View of Hindoostan*, ii).
27. Thomson seems to have consulted numerous travel narratives for *The Seasons*, including Harris's *Collection of Voyages* (1705), Raleigh's *Discovery of Guiana* (1596), and Churchill's *Collection of Voyages* (1732), among other; see McKillop, *Background of Thomson's* Seasons, 159, 165.
28. Pennant, *Tour in Scotland*, ii.
29. Barratt explains that translations of Thomson's *Seasons* published in Russia from the 1770s through the early nineteenth century strongly influenced that nation's literary turn toward sensibility ("James Thomson in Russia," 367–73).
30. Without referencing Thomson, Sherry recently used the phrase "backward time" to discuss fin de siècle decadence and an adverse, past-directed orientation in modernism (*Modernism and the Reinvention of Decadence*, 207).
31. As Sachs extensively argues, "Anxieties about decline—national and imperial, economic and political, cultural and literary—are a pervasive feature of British public discourse in the later eighteenth and early nineteenth century" (*Poetics of Decline*, 4).
32. Barrell, *Poetry, Language, and Politics*, 102.
33. According to Wilson, "People from frigid and torrid climes had the same physical and moral weaknesses. In these cases, climatic factors had to be invoked to explain their *backwardness*" (quoted in Golinkski, *British Weather*, 182; emphasis mine).
34. Palmeri, *State of Nature*, 5. For a study of women writers' interactions with these ideas, see DeLucia, *Feminine Enlightenment*.
35. Palmeri, 41.
36. See, for example, various essays in Fox, Porter, and Wokler, eds, *Inventing Human Science*.
37. Palmeri, *State of Nature*, 41–42.
38. Palmeri, 31.
39. Many additional poets who did not employ this title nevertheless incorporated its themes into their verses, as did Thomas Warton, for instance, in his "Ode on the First of April" (1777).
40. Golinski, *British Weather and British Romanticism*, 4, 92.
41. While I argue that this is the case for these poets writing verses on the backwardness of spring, it is worth noting that some other poets,

including Thomas Warton in *The Pleasures of Melancholy* (1747), more positively associate winter with the personifications of Melancholy and Contemplation.

42. West, "Ode to Mr. Gray," in *Poems of Mr. Gray*, 147–48. All references to West's poem are cited parenthetically by line number.
43. See Thomson's notes to *Summer*, ll. 738–44.
44. Barbauld, *Poems*, 243. All references to Barbauld's poem will be cited parenthetically by line number. Barbauld was certainly familiar with *The Seasons*, and she drew one of her stories in *Lessons* from a passage in *Autumn*; see McCarthy, *Anna Letitia Barbauld*, 202.
45. In some traditions, it is four (not six) pomegranate seeds, and thus months.
46. Barbauld, *Lessons for Children*.
47. Aikin, *Calendar of Nature*, 14.
48. Penn, *Poetical Miscellanies*, 74–75. All references to Penn's poem will be cited parenthetically by line number.
49. Bowden, ed., *Eighteenth-Century Modernizations*, 244.
50. Mossner, *Life of David Hume*, , 67.
51. See Bailes, "Literary Plagiarism and Scientific Originality," 275, 276.
52. Roger, *Buffon*, 263.
53. This poem is quoted in full in George's *Botany, Sexuality, and Women's Writing*, 189–93.
54. George, "Not Strictly Proper," 205n1.
55. Shteir, *Cultivating Women*, 240n4; George, "Not Strictly Proper," 191; Haut, "Reading Flora," 246.
56. Bailes, *Questioning Nature*, 85.
57. This is the first stanza of a poem that can be found attributed to William Hayley and quoted in Seward, *Letters*, 23.
58. For example, two poems referencing Seward's name in this way, by T. Park and H. F. Carey, are printed at the beginning of Seward, *Original Sonnets on Various Subjects*.
59. King-Hele, *Erasmus Darwin: A Life*, 117.
60. Seward, *Memoirs*, 131.
61. For more on the Lichfield literary circle, see, for example, King-Hele, *Erasmus Darwin: A Life*, 117–18.
62. Mundy, *Needwood Forest*. All references to Mundy's poem are from this edition and cited parenthetically by page (prose) or line number (verse). The published version of Mundy's "Backwardness" poem differs from the anonymous manuscript transcription in that it includes about twenty additional lines and minor word variations. Many thanks to Allison Davis for alerting me to this edition.
63. Barnard, "Anna Seward's Hidden Words," 427. According to Barnard's assessment, Darwin's and Seward's additional stanzas seem to constitute an

unsuccessful instance of literary borrowing as these lines are clearly "no longer Mundy's voice" and "fragment" his work (427, 428).
64. When Mundy published *Needwood Forest*, Darwin, Seward, and Boothby each wrote poems appended to *Needwood* celebrating Mundy's verses and invoking his name, again displaying that this sort of poetic lauding of one another, as Mundy does for Seward in his "Backwardness of Spring," was very common within the Lichfield literary circle. Mundy, *Needwood Forest*.
65. Mahood, *Poet as Botanist*, 53n12.
66. Martin Rowley researched this weather data, which can be found on the Weatherwebdotnet website: https://premium.weatherweb.net/weather-in-history-1750-to-1799-ad/ (accessed 26 July 2022). According to these records, the year 1772 appears not to have had a backward spring, again displaying that this date was an error in the transcription of that manuscript version of the poem.
67. George, "Not Strictly Proper," 191. This speech was delivered in December 1772, and thus after the spring of that year, adding further evidence that the spring of 1772 was not that described by the poem; however, since the poem is thought to have been penned in the 1780s, it is possible that the author was influenced by the speech regardless of this chronology.
68. King-Hele, *Erasmus Darwin and the Romantic Poets*, 25; see also Mahood, *Poet as Botanist*, 53.
69. Priestman, "Progress of Society?," 311. Interestingly, Darwin also references *Fable of the Bees* in his text *Zoonomia*, 1:433.
70. Priestman, "Progress of Society?," 310.
71. Darwin, *Temple of Nature*, 19.
72. For a more thorough treatment of Buffon's theory, see Oldroyd, *Thinking about the Earth*, 91.
73. Darwin, *Botanic Garden*, 51.
74. Darwin, 51.
75. Darwin, 51. For Darwin's extensive, scientific interest in meteorology, see King-Hele, *Erasmus Darwin and the Romantic Poets*, 20.
76. Darwin, *Economy of Vegetation*, canto 1, ll. 434–40.
77. Montesquieu, *Spirit of the Laws*; for some of the ideas and sentences in this paragraph, see also Bailes, *Questioning Nature*, 139.
78. Bate, *John Clare*, 89.
79. Bate, 89.
80. McKusick, "Beyond the Visionary Company," 224; Bate, 90.
81. Clare, "Backward Spring," l. 3, in *Rural Muse*, 104. All references will be cited parenthetically by line number.
82. Burnet, *Sacred Theory of the Earth*, 1:241, 95.
83. Burnet, 1:243. As McKillop notes, "Thomson knew well enough that Burnet was to be put among the 'fabling poets,' yet it is clear that he was fascinated by"

this theory (100). Golinski explains that the seasons "were widely regarded as an aspect of God's providential design of the cosmos. Newton himself thought it plausible that God had tilted the earth's axis in relation to the plane of the ecliptic in order to cause them" (*Background of Thomson's Seasons*, 96).
84. Burnet, *Sacred Theory of the Earth*, 1:244.
85. As King-Hele notes, "The three-year average for 1784–6 is the lowest on record, a mini ice-age created by the volcano in Iceland" (*Erasmus Darwin: A Life*, 208). King-Hele here alludes to the Laki eruption in southern Iceland; the 1815 volcanic eruption refers to Mount Tambora in Indonesia, extensively explored, for instance, in Wood, *Tambora*.

2. Linnaeus's Botanical Clocks

1. Cox, *Descriptive Catalogue*, 25.
2. Cox, 31–32.
3. For instance, Percy Shelley's fleeting "flower that smiles today / Tomorrow dies" contrasts with William Blake's time-transcending "Wild Flower" that promises "eternity in an hour."
4. Dunlap, Loros, and De Coursey, eds., *Chronobiology*, xvii.
5. DeCoursey, "Behavioral Ecology and Evolution," 42.
6. Some of these botanists include Henri-Louis Duhamel de Monceau, *La Physique des Arbres* (Paris: H. L. Guerin and L.F. Delatour, 1758), Johann Gottfried Zinn, and, in the early nineteenth century, Augustus Pyramus de Candolle, *Physiologie Vegetale* (Paris: n.p., 1832).
7. DeCoursey, "Behavioral Ecology," 42.
8. Information in caption to fig. 3 from Somers, "Physiology and Molecular Bases," 10. According to Oxley, the dial in figure 3 was painted by Ursula Schleicher-Benz in 1948 (*Botanical Illustration*). A more contemporary to these poems, though slightly less decorative, image of a Linnaean "botanical clock" can be found, for instance, in Taylor's *Complete Weather Guide*.
9. Grainger, *Sugar-Cane*, 146.
10. Browne, "Botany for Gentlemen."
11. Koerner, *Linnaeus*, 22, 48, 55.
12. Gibson, *Animal, Vegetable, Mineral?*, 154.
13. Thomas Percival founded the Manchester Literary and Philosophical Society in 1781 and was president of the society until his death in 1804. Darwin gained honorary membership of the Manchester Society in 1784, see King-Hele, ed., *Letters of Erasmus Darwin*, 138. Percival also corresponded with other figures in Darwin's circle, such as Anna Seward.
14. Reill, *Vitalizing Nature*, 34, 48. See also Gaukroger, *Collapse of Mechanism*, 344.
15. Packham, "Science and Poetry of Animation," 191–92.

16. Southey, "Collected Works," 71.
17. Barbauld, *Pleasures of Imagination*, 4.
18. Jackson, "Rhyme and Reason," 176, 181, 194; Packham, "Science and Poetry of Animation," 200. To recount further critical perspectives on this subject, Hassler states that Darwin read Boerhaave, Haller, and David Hartley and calls him a mechanist and "a materialist of the extreme La Mettrie type," referencing Julien Offray de La Mettre, author of *L'Homme Machine* (1748); Hassler, *Comedian as the Letter D*, vii, 24–25. In contrast, Delaporte draws the vitalism mechanism divide in botany as one between "plant action theory," with which he associates the use of analogy in the works of Charles Bonnet and Erasmus Darwin, and "plant mechanism," with which he associates experimentalism and Cartesian mechanism in the works of John Lindsay and Thomas Knight, and suggests that the adherent to analogy and plant action theory "went wrong" (*Nature's Second Kingdom*, 150).
19. Reill, *Vitalizing Nature*, 12. Griffiths also posits this possibility of applying Reill's conception of Enlightenment vitalism to Erasmus Darwin's works in "Intuitions of Analogy," 660.
20. In addition to Packham's "Science and Poetry of Animation" and Griffiths's "Intuitions of Analogy," on Darwin's use of analogy see also Porter, "Scientific Analogy and Literary Taxonomy." In the Lunar Society of Birmingham, Darwin maintained friendships not only with other naturalists but also with many leaders of the industrial revolution, including John Whitehurst, an "immensely skillful . . . master of clocks," and he invented several mechanical devices, such as a small mechanical bird, a copying machine, a steering mechanism for his carriage, and a speaking machine (King-Hele, *Erasmus Darwin: A Life*, 38). For more on the Lunar Society, see Uglow, *Lunar Men*.
21. McNeil, *Under the Banner of Science*, 153.
22. Darwin, *Loves of the Plants*, 62–63.
23. Dunlap, *Chronobiology*, 20.
24. In this vein, "circadian or circannual rhythms are exclusively those that match an environmental periodicity but have been shown to be generated by an underlying endogenous pacemaker system" (Dunlap, 29).
25. William Blake also references this vegetable motion in his "Song of Experience" that begins "Ah Sun-flower! weary of time, / Who countest the steps of the Sun" (ll. 1–2).
26. Milne, *Botanical Dictionary*, 283.
27. Linnaeus, *Philosophia Botanica*, 296–97.
28. Attentive to economic implications of natural history, Linnaeus advocated "transmutationist botany, a science that assumed that nature was so malleable that by means of floral transplants naturalists could assure independent

yet complete state economies" (Koerner, *Linnaeus*, 108). The floral clock thus imparts, within a single concept, his sense of ordered, mechanized nature, in which flowers contain practical use value and multiple means of contributing to national economies, necessitating heightened research and valuing of this science.

29. Pliny the Elder. *Natural History*, 5.349.
30. Grene and Depew, *Philosophy of Biology*, 37.
31. Cole and Markley, "Human, Animal, and Machine," 376.
32. Perhaps the most famous statement of this teleological "argument from design" employing the watchmaker analogy was in William Paley's *Natural Theology, or Evidences of the Existence and Attributes of the Deity* (1802). While Darwin sometimes pays lip service to the existence of a Creator, his contemporaries often accused him of atheism.
33. Priestman identifies Darwin as expressing "nominally deistic materialism" and, analyzing this particular passage, remarks, "By a common Enlightenment ambiguity, the 'Superstition' to be attacked for its overweening power and abused wealth might either be read as a safely distant Catholic Church, or as established religion more generally. Whichever is the case, its destruction by Time is inevitable" (*Romantic Atheism*, 49, 63).
34. Interestingly, although Linnaeus "believed that he was one of the elect, called upon by God to reveal, Moses-like, the divine law of nature," he likewise limited the afterlife to the production of blood-descendants (Koerner, *Linnaeus*, 23, 89).
35. See, for instance, *Phytologia*, where Darwin states, "Sexual generation [is] the chef d'oeuvre of nature" (81). Schiebinger writes, "For [Erasmus] Darwin, sex was not just the mechanism for improving and diversifying the stock of living organisms: it was also the purest source of happiness" (*Nature's Body*, 31).
36. DeCoursey, "Behavioral Ecology," 42.
37. In this poem, Darwin arguably presents even the Sensitive Plant's movements largely as passive reactions to external forces, likening its motions to those of a thermometer or compass (*Loves of the Plants*, ll. 299–314).
38. Thornton, *New Illustration*, 15.
39. Darwin, *Phytologia*, 119–26; as Darwin states, he began exploring these ideas in his *Zoonomia* (1794), and he gestures toward them in the *Botanic Garden* as well. Darwin also endowed plants with "irritability," a catchword introduced by the mechanist Albrecht von Haller that often helped support antimechanistic tendencies and thus pointed to the strained relation between mechanism and concepts of organic nature in late eighteenth-century biology (*Phytologia* 2, 1; *Philosophy of Biology*, 63).
40. Darwin, *Phytologia*, 80, 83.

41. Whippo and Hangarter, "'Sensational' Power of Movement," 2118. According to Whippo and Hangarter, Darwin "believed that plant movement, like animal behavior, would eventually be explained in materialistic terms" (2118).
42. See Gibson, *Animal, Vegetable, Mineral?*, 166.
43. Some essays addressing Charlotte Smith's writings in relation to botany include Pascoe, "Female Botanists"; Ruwe, "Charlotte Smith's Sublime"; Landry, "Green Languages?"; Dolan, *Seeing Suffering*; and Kerr, "Melancholy Botany."
44. George, *Botany*, 106.
45. Linnaeus, *Elements of Botany*, 396.
46. Reill, *Vitalizing Nature*, 179.
47. Perovic, *Calendar in Revolutionary France*, 3.
48. Hilbish, *Charlotte Smith*, 211; emphasis mine.
49. Regarding social class distinctions, Delaporte explains that the floral clock "itself might be classed, along with the garden motifs, in the category of 'barrochus rupestris', as Eugenio d'Ors notes that 'the essence of the baroque always contains something of the peasant'" (*Nature's Second Kingdom*, 183).
50. His poetic descriptions, combined with the naturalist's "antirhetorical stance, and his appeals to virtue ... struck a chord in Europe's greatest men of letters" (Koerner, *Linnaeus*, 25).
51. See Knight, *Ordering the World*, 23.
52. Knight, 63.
53. Darwin, *Phytologia*, 512.
54. Smith, for instance, undermines the botanist's reliability in her note on the *Fly Orchid* in "Beachy Head," stating that "Linnaeus, misled by the variations to which some of this tribe are really subject, has perhaps too rashly esteemed all those which resemble insects, as forming only one species."
55. Cloudsley-Thompson, *Biological Clocks*, 22.
56. Milne, *Botanical Dictionary*, 443.
57. Milne, 284.
58. Whereas the mechanical clock appeals aurally and visually, the flowers indulge sight and smell. In fact, some contemporaries suggested configuring a floral clock either according to scent or to color changes certain flowers undergo over the course of the day (Shoberl, *Language of Flowers*, 356).
59. Milne, *Botanical Dictionary*, 444.
60. Yolton, *Thinking Matter*, 32; Drury, "Haywood's Thinking Machines," 209.
61. Shelley, *Sensitive-Plant*, ll. 101, 103–5. Mitchell calls Percy Shelley an experimental vitalist (*Experimental Life*, 11). Kelley analyzes Percy Shelley's *Sensitive-Plant* and applies to it "the phrase *embodied life* to refer to material forms of life without claiming that those forms can be separated from

mechanical or nonvital processes," for "romantic and postromantic efforts to bifurcate vital and mechanical processes neglect what experimentalists and poets found more plausible after 1800—the possibility that forms of life and nonlife, as well as vital and mechanical operations, coexist in forms of being we call life" (*Clandestine Marriage*, 210).

62. Locke, *Essay Concerning Human Understanding*, 270.
63. Coleridge, of course, framed his organic formulation of the imagination against "mechanical philosophy" and in terms of rampant plant metaphors, declaring to Wordsworth in 1815 that philosophy must embrace "life and intelligence . . . considered in its different powers from the plant up to" humanity, and turn away from "the philosophy of mechanism, which, in everything that is most worthy of the human intellect, strikes Death, and cheats itself by mistaking clear images for distinct conceptions" (*Letters*, 2:649; Locke, 267–68).
64. *Monthly Review* 56 (May–August 1808): 99–100.
65. *British Critic* 30 (1807): 170, 174.
66. See, for instance, Charles Lamb's letter to Coleridge, "damn[ing]" Anna Barbauld for integrating natural history with imaginative children's works (Marrs Jr., *Letters*, 2:82).
67. In 1828, with the death of Sir James E. Smith, founder and president of the Linnean Society, and the appointment of John Lindley, a strong advocate of the natural system, as the first professor of botany at London University, "'scientific' botanists were turning . . . from taxonomy to physiology and morphology, and from the Linnaean sexual system to the natural system of classification" (Shteir, *Cultivating Women*, 155).
68. Gibson, *Animal, Vegetable, Mineral?*, 168–69.
69. Knight, *Ordering the World*, 118. Kelley explains that "Linnaean systematics dominated botanical practice from about 1750 to 1810 and, among amateurs and field naturalists, long after that," while "the so-called Natural System gained prominence in Europe in 1789, when Antoine-Laurent de Jussieu published *Genera Plantarum*" (*Clandestine Marriage*, 19, 20).
70. Gray, *Natural Arrangement*, 1:22.
71. John Clare's journal entry for 24 October 1824, *Prose of John Clare*, 117.
72. Hemans later included "The Dial of Flowers" in the second edition of her volume *The Forest Sanctuary, with Other Poems* (1829).
73. Williams, *Literary Women of England*, 492.
74. Kelly, ed., *Felicia Hemans*, 22.
75. See, for instance, Leighton, *Victorian Women*.
76. *Monthly Repository and Review*, 924.
77. King-Hele, *Erasmus Darwin: A Life*, ix; Cloudsley-Thompson, *Biological Clocks*, 10.
78. Linnaeus, *Philosophia Botanica*, x; italicization of "cup" and "bell" mine.

79. Milne, *Botanical Dictionary*, 284.
80. Darwin, *Phytologia*, 314.
81. For a discussion of this shift in critical perceptions creating the mythification of "high" literature in this era, see Macfarlane, *Original Copy*.
82. See Melnyk, "William Wordsworth and Felicia Hemans," 144.

3. Transformations of Gender, Race, and Poetic Sensibility

1. *OED Online*, s.v. "trans-, prefix," http://www.oed.com (accessed 28 July 2022).
2. Buffon explained this hypothesis in his section titled "Of the Degeneration of Animals" (1766), which may be found in vol. 7 of *Natural History*. References are to this edition and will be cited parenthetically. For more on Buffon's theory, see Egerton, *Roots of Ecology*, 85–86; Parrish, "Science, Nature, Race," especially 476; and Sloan, "Gaze of History," 122–23, 135–38.
3. Riddell, *Voyages to the Madeira*. All references to this work will be cited parenthetically.
4. For more on Riddell and concepts of national hybridity, see Bailes, *Questioning Nature*, 47–70.
5. See Riddell's biographers, Macnaghten, *Burns's Mrs. Riddell*; and Gladstone, *Maria Riddell*, 5.
6. Kerr, *Memoirs*, 2:387. References are to this edition and will be cited parenthetically.
7. She describes meeting James Boswell, the biographer of Samuel Johnson, and quips that "a stranger biped, yourself always excepted, I know no where." The exhortation to focus on "human life" echoes that of Johnson in his essay on travel writing in *Idler*, no. 97 (23 February 1760).
8. Wollstonecraft, *Letters*. See also Polwhele's portrayal of Wollstonecraft in *Unsex'd Females*. As Juengel notes, "Wollstonecraft consistently invokes abolitionist rhetoric in her critique of patriarchal hegemony, exploiting the affective force of juxtaposing the domesticated woman and the subjugated African" ("Countenancing History," 899).
9. Gladstone, *Maria Riddell*, 31, 29.
10. Gladstone, 27.
11. See also Gladstone, 37.
12. Walpole to More, 26 January 1795, in *Letters*, vol. 15, letter 2956. See also Bewell, "Hyena Trouble."
13. See Bailes, "Hybrid Britons," 212.
14. Darwin, *Botanic Garden*, 7.
15. Darwin also was interested in zoology, and in his text *Zoonomia* (1794–96), where he "attempt[s] to classify animal orders, he emphasizes primal desires

(lust, hunger, and security) as the stimuli of transformation." Teute, "Loves of the Plants," 329.
16. Mills, *Biopolitics*, 158.
17. Smellie, *Philosophy of Natural History*, 247; emphasis mine. References are to this edition and will be cited parenthetically.
18. Casid, *Sowing Empire*, 21.
19. Casid, 2.
20. For the broader context of this idea, see Nussbaum, *Limits of the Human*, 11–12; and Nussbaum, *Torrid Zones*.
21. Long, *Candid Reflections*; quoted in Wheeler, *Complexion of*, 141.
22. Long, *History of Jamaica*, 2.2.13.336, 2.2.13.327.
23. One might think about the implications for class in the historical context of the French Revolution, since Smellie's *Philosophy* was published in 1790.
24. Casid, *Sowing Empire*, 1.
25. See, for instance, Bewell, "Jacobin Plants"; Shteir, *Cultivating Women*; and George, *Botany*.
26. As Thompson astutely notes, "One wonders how far such flirtation was for [Riddell] a social performance necessary for retaining a useful scientific contact. There are several moments in these letters, for example, where Riddell gently chides Smellie for over-indulging in flights of fancy at the expense of sending on more useful information, such as 'any additional acquisitions in the vegetable way' at the Edinburgh Botanic Garden" ("Women Travelers," 441).
27. Interestingly, James Thomson was given the sinecure of surveyor-general of the Leeward Islands from 1744 to 1746. Thomson, *Seasons and the Castle of Indolence*, xxi.
28. Schaw, *Journal*, 102. References are to this edition and will be cited parenthetically.
29. As Koerner has shown, the Linnaean system also related to resource extraction and global capital, so that Linnaeus's "natural science underwrote his political economy" ("Purposes of Linnaean Travel," 119).
30. Krueger, *Transatlantic Women Travelers*, 2.
31. Krueger, 2.
32. Johnson, "Thresholds of Livability," 50.
33. See Coleman, "Janet Schaw and the Complexions of Empire," 171–72.
34. Coleman, 172.
35. Chiles, *Transformable Race*, 2.
36. On Buffon's thoughts about racial transformation and monogenesis, see Curran, *Anatomy of Blackness*, 74–116.
37. Goldsmith, *History of the Earth*, 2:242.
38. Merian, *Metamorphosis Insectorum Surinamensium*, 40; Schiebinger, *Plants and Empire*, 1–3. On uses of plants by European, Indigenous, and African

medical practitioners in European colonies, see Schiebinger, "Scientific Exchange"; and Weaver, *Medical Revolutionaries*.

39. In his footnote to *The Sugar-Cane, A Poem* that describes women's use of cashew nut oil, James Grainger remarks on the pain of this freckle-removing process. Riddell's description of the process as "excruciating" does not necessarily indicate her participation; however, the physiological intensity of the term "excruciating," combined with her involvement in West Indian society, her personally testing the uses of many island resources, and her mother's potential hailing from St. Kitts open the possibility that Riddell may have tried this exfoliating process.
40. Schaw, *Journal*, 114.
41. Schaw, 114.
42. Schaw, 115.
43. Darwin, *Zoonomia*, 2:375. All references to this work will be cited parenthetically.
44. Clare, *Shepherd's Calendar*, 48.
45. Green, *Universal Herbal*, 585.
46. Browne, *Civil and Natural History*, 227.
47. Riddell also writes of this poisonous fluid that "the Caribs and Indians dip the points of their arrows in it when they wish the wounds they inflict to prove mortal" (*Voyages*, 91).
48. As Wilson points out, to many eighteenth-century Britons, it appeared that "the national character, acquired through propinquity, could when removed from the structures of civilized life quickly give way" (*Island Race*, 13). Wheeler likewise relates that, in the Caribbean, "many observers concluded that Englishness showed alarming evidence of degeneration and vulnerability" (*Old Enemies*, 226).
49. See also Kim, "Complicating 'Complicity/Resistance,'" 172–77. On Schaw's views of slavery, see Bohls, *Slavery and the Politics of Place*, 143–64.
50. Coleman, "Janet Schaw," 177. Krueger rightly cautions "that the freedoms afforded to some women travelers in this era, especially those of white European descent, were the result of imperialism, colonization, and Black women's trauma" (*Transatlantic Women Travelers*, 2).
51. Long, *History of Jamaica*, 2.13.327. Schmidgen discusses more positive perspectives of racial hybridity within English identity in the seventeenth century, and points out that, at that time, "Ancient Rome's policy of mixture with the conquered" was familiar "as one of the keys to imperial expansion and integration" (*Exquisite Mixture*, 150).
52. Petley, "Gluttony, Excess," 90–91. On the changing British perceptions of their West Indian colonies during these decades, see Greene, *Evaluating Empire and Confronting Colonialism*.

53. See Bailes, "Hybrid Britons," 214–15.
54. Parrish, *American Curiosity*, 1, 235, 251. Parrish discusses fears held by white colonists that they could be poisoned by enslaved Africans, who had knowledge of local plants (274). As Fulford, Lee, and Kitson show, examination of native use of plants could sometimes be a detriment to enslaved populations and the poor. For example, while Indigenous peoples of the Pacific Islands ate breadfruit as a staple, when British colonizers exported the plant to the West Indies as a cheap food source, it was said that the enslaved workers hated this foodstuff (*Literature, Science, and Exploration*, 117).
55. On enslaved Africans' gardens and productions through agriculture and husbandry, see Tobin, *Colonizing Nature*, 56–80.
56. Livesay notes that Edmund Burke's 1780 "Sketch of a Negro Code" suggests "that the slave system could be improved if black laborers married and lived in families modeled after English households," a growing sentiment that Livesay argues made "the prospect of interracial sex" become "deeply problematic" ("Decline of Jamaica's Interracial Households," 113).
57. See Chiles, *Transformable Race*, 6. Curran also examines Buffon's response to this portrait ("I Altered," 102–5).
58. OED Online, s.v. "piebald," http://www.oed.com (accessed 28 July 2022).
59. Sweet, *Bodies Politic*, 276.
60. Sweet, 276–77.
61. Wilson, *Island Race*, 14.
62. Wheeler, *Complexion of Race*, 299.
63. Braunschneider writes, "Eighteenth-century philosophical and scientific arguments often suggest that some sorts of people, usually Africans, are more like animals than others . . . frequently such comparisons served to justify the exploitation of humans" ("Lady and the Lapdog," 47).
64. Macnaghten, *Burns' Mrs. Riddell*, 82.
65. Quoted in Macnaghten, 82.
66. Gladstone, *Maria Riddell*, 31.
67. Quoted in Macnaghten, *Burns' Mrs. Riddell*, 101. Unfortunately, Burns's gravestone, as Riddell envisioned it, never came into being.
68. Gladstone, *Maria Riddell*, 87.
69. Riddell, ed., *Metrical Miscellany*.
70. *British Critic* 20 (July–December 1802): 258–61. The review reveals or acknowledges Riddell as the editor. Additionally, it was in 1802 that Riddell received word of her husband's death in Antigua. Soon after, Riddell and her daughter took apartments in Hampton Court Palace, where she continued to receive regular visits from literary-minded friends.
71. King-Hele, *Erasmus Darwin: A Life*, 125–26.
72. King-Hele, 136.

73. King-Hele, 156.
74. King-Hele, 62.

4. Cultivated for Consumption

1. Geoghegan, *Irish Act of Union*, vii.
2. Geoghegan, vii.
3. Kiberd, *Inventing Ireland*, 251.
4. Kirkpatrick, "Introduction," vii.
5. Trumpener, for instance, calls *The Wild Irish Girl* an "allegorical presentation of the contrast, attraction, and union between disparate cultural worlds" ("National Character, Nationalist Plots," 697). While Kirkpatrick views Owenson as "defiant" in "seek[ing] to provide a genealogy for a separate Irish identity at a historical moment when that identity seemed lost" in the Act of Union ("Introduction," vii), Moore "accuse[s]" Owenson of using her novel to "support" the Union ("Acts of Union," 114), Nersessian portrays it as "a point of convergence between real and ideal conditions, between historical time and the imagination of some better future" ("Empire and Attachment," 340), and Tracy lauds it as "a re-imagined distribution of power between Britain and Ireland, as well as between men and women" ("Mild Irish Girl," 82).
6. For good general studies of ties between imperialism and botany, see, for example, Grove, *Green Imperialism*; Schiebinger, *Plants and Empire*; and Casid, *Sowing Empire*.
7. Owenson, *Wild Irish Girl*, 1. All references to this work will be cited parenthetically. Lew, "Sidney Owenson," 44.
8. From Fazio Dagli Uberti's *Il Dittamondo* (*The Book of the World*), published in 1346. Uberti was born in Pisa and wrote a fictional journey around the world.
9. Brown, "Hunger," 56.
10. For a broader discussion about forms of Irish consumption, see Powell, *Politics of Consumption*.
11. Belanger, *Critical Receptions*, 7.
12. Campbell, *Lady Morgan*, 2.
13. Spenser, *View of the Present State*, 6:154.
14. Rawdon, "Particulars," 93. My thanks to Noah Heringman for alerting me to this article.
15. Rawdon, 92; Moryson, *History of Ireland*, 2:282–84.
16. Rawdon, 92.
17. Swift, *Modest Proposal*, 7, 15.
18. Kirkpatrick, "Introduction," xvi; Avramescu, *Intellectual History*, 45. Owenson defended the actions of the French peasantry, declaring, "the total

overthrow of that frightful system of feudality which had so long crushed them into slavery, was among the first and best works of the revolution." See Donovan, *Sydney Owenson*, 139–40.

19. Gigante, *Taste*, 119. See also Wilson, *Island Race*, 91.
20. De Nie, *Eternal Paddy*, 56.
21. Quoted in Avramescu, *Intellectual History*, 172.
22. Rawdon, "Particulars," 92. Also, as Donovan points out, Owenson "probably extracted the Uberti quotation" about Irish cannibalism discussed at the beginning of this chapter, "from the Earl of Charlemont's 1787 essay, 'The Antiquity of the Woollen Manufacture in Ireland,'" and this Earl of Charlemont was the son of the Countess of Moira ("Text and Textile," 40).
23. One example of a contemporary text specifically devoted to British botany is William Withering's *Botanical Arrangement of All the Vegetables Naturally Growing in Great Britain*. Modern critical studies of these comparisons between national character and natural objects include Bewell, "Jefferson's Thermometer"; and, more particularly to botany, George, *Botany*.
24. Darwin, "Proem," in *Botanic Garden*, v–vii; Darwin, *Phytologia*. Heckendorn Cook discusses Erasmus Darwin's engagement with phytodynamism—that is, vegetal mobility—in "'Perfect' Flowers, Monstrous Women," 252.
25. Burn, "Second Address"; quoted in Plasa, *Slaves to Sweetness*, 44. See also Sussman, *Consuming Anxieties*, 14, 115.
26. Owenson, review of J. G. C. Feuillide's *Ireland*, 273; quoted in Donovan, *Sydney Owenson*, 188.
27. Rodgers, *Ireland, Slavery, and Anti-Slavery*, 2.
28. Salaman, *Social History and Influence of the Potato*, 453, 486. Hall also discusses potatoes in the context of Owenson's novel ("Wild Irish Girl Diet").
29. Groom, "William Henry Ireland," 29.
30. Groom, 28.
31. Of course, the most famous artificial system of plant taxonomy in the eighteenth century was that of Carl Linnaeus, which classified according to the single character of plants' sexual parts.
32. In this era, the English also were sometimes identified with particular plants. For example, Sam George discusses comparisons of English character with the English Rose, which was sometimes revered in opposition to gaudy tulips and their associations with the Dutch (*Botany*, 156).
33. White, *Natural History of Selborne*, 93.
34. See, for instance, Montesquieu, *Spirit of the Laws*.
35. Polwhele, *Unsex'd Females*, 8. See also Schiebinger, *Nature's Body*, 1–30; and Shteir, *Cultivating Women*.
36. Owenson, *Lay*.
37. Owenson, 39.
38. Owenson, 87; she cites Lobb, *Contemplative Philosopher*, 1:252.

39. There is also arguably a class distinction implied in Horatio's attempt to infuse Glorvina with cosmopolitan cultural interests while the Irish lower classes are correlated with primarily Irish ideas and natural objects. Turner explains that in the late eighteenth century the lower and middle classes increasingly "claimed most insistently to embody Englishness or Britishness, in contrast to the unpatriotic cosmopolitanism of the aristocracy" (*British Travel Writers in Europe*, 17). As I discuss elsewhere, Oliver Goldsmith asserted that the lower classes of every nation most strongly exhibit their country's national characteristics or identity (Bailes, "Literary Plagiarism and Scientific Originality," 269). Thus, botanically, whereas Owenson associates the Irish peasantry with Irish vegetation, she signifies the developing marital union between England and Ireland with flowers that are either equally familiar or equally exotic to both nations.
40. Some examples of these authors' contributions to botanical thought include: Jacques-Henri Bernardin de Saint-Pierre, *A Voyage to the Island of Mauritius*; Jean-Jacques Rousseau, *Letter on the Elements of Botany, Addressed to a Lady*; and Johann Wolfgang von Goethe, *Versuch die Metamorphose der Pflanzen zu erklaren* [*Metamorphosis of Plants*]. Additionally, Chateaubriand's *Atala* incorporates some of his studies and observations of the flora in North America.
41. Montagu, *Letters*, 3:5. See also Seaton, *Language of Flowers*.
42. Tessone, "Displaying Ireland," 170.
43. Ferris, "Narrating Cultural Encounter," 298.
44. Bhabha discusses the "potential political subversion in [the] strategy of the evil eye," the "petrifying, unblinking gaze that falls Medusa-like on its victims" (*Location of Culture*, 80).
45. Pratt, *Imperial Eyes*, 56.
46. Pratt, 57.
47. Pratt, 57.
48. Kirkpatrick, "Introduction," xiii, xiv, xvi–vii.
49. Colley, *Britons*, 5. For more on the relationship between Catholics and Protestants in this era, see, for example, Wheeler, *Old Enemies*.
50. Gigante, *Taste*, 27.
51. Tracy, "Mild Irish Girl," 88.
52. For discussions of hybridity in the eighteenth and nineteenth centuries, see, for example, Schmidgen, *Exquisite Mixture*; and Young, *Colonial Desire*. The kind of Irish hybridity depicted by Owenson arguably aligns more closely with Bakhtin's notion of "organic hybridity," forming an amalgamation or "creolization," rather than with the contestatory potential in his view of "intentional hybridity." According to Bakhtin, authoritative discourse "is by its very nature incapable of being double-voiced; it cannot enter into hybrid constructions" without immediately undermining its authority (*Dialogic Imagination*, 344).

53. See George, *Botany*, 162.
54. George, *Botany*, 32.
55. Braun, "Seductive Masquerade," 36.
56. Ferris, "Narrating," 291.
57. Late in his career, Linnaeus offered challenges to this theory of botanical hybridity; nevertheless, this perception prevailed in his works and in those of many other naturalists of the era. Koerner, *Linnaeus*, 44; Bowler, *Evolution*, 70.
58. Bhabha suggests that intentional hybridity may function as challenge or resistance by depriving "the imposed imperialist culture, not only of the authority that it has for so long imposed politically, often through violence, but even of its own claims to authenticity" ("Postcolonial Critic," 48). In Glorvina's education of Horatio, she may be seen as achieving a form of this resistance, but the novel also indicates that the goal of English imperial efforts in Ireland is to subsume Irish culture, and it is Irish culture and "authenticity" that seems most "challenged" by the potential hybridity of this union.
59. Although, in the text, Lord M. states that the Irish are not "creatures of the soil," that is precisely how he goes on to portray them, as plants (250).
60. As Drayton writes, "If Christian Providentialism was the ideological taproot of the 'First British Empire,' Enlightenment doctrines of cosmopolitan progress, anglicized in the Georgian idea of 'Improvement,' sustained its successors" and "what distinguishes British imperialism, from the late eighteenth century onwards, is this faith in its capacity and right to increase the happiness of barbarians" (*Nature's Government*, 93).
61. Lew, "Sidney Owenson," 41.
62. Quoted in Newcomer, *Lady Morgan*, 33.

5. "On the Green Margin"

1. Smith, *Conversations*, lines 10, 13, 62, 66. Unless otherwise specified, all quotations of Charlotte Smith's poetry are from *Poems*. All quotations from "Flora" will be cited parenthetically by line number, and quotations from other poems will generally be cited parenthetically by page number.
2. Smith, *Works*, 13:226.
3. For excellent background information on botany's role in women's education in the Romantic period, see Shteir, *Cultivating Women*; and Page and Smith, *Women, Literature, and the Domesticated Landscape*.
4. Dolan, *Seeing Suffering*, 101–31.
5. Andrews, "Charlotte Smith," 1158.
6. Fletcher, *Charlotte Smith*, 239. In addition to this reference regarding Smith's drawing of flowers, as well as in her related sonnet "To Dr. Parry of Bath, with Some Botanic Drawings Which Had Been Made Some Years," she

also describes her painting of them in her poem "Sent to the Honorable Mrs. O'Neill, with Painted Flowers."
7. Fletcher, *Charlotte Smith*, 323, 329.
8. Smith, *Works*, 13:227.
9. Fletcher, *Charlotte Smith*, 12.
10. Fletcher, 13.
11. Darwin, *Loves of the Plants*, canto 4, l. 270n.
12. OED definitions of "margin," www.oed.com (accessed 16 August 2022).
13. Citing Julie Ellison, Porter notes that "fancy is both an aesthetic and an empirical mode, one that legitimates the poet's scientific knowledge" ("From Nosegay to Specimen Cabinet," 41).
14. Shteir further discusses traditions of flora and Darwin's Goddess of Botany in her essay, "She comes!—the Goddess!"
15. Darwin, *Loves of the Plants*, vi.
16. Mahood, *Poet as Botanist*, 20; Pascoe, "Female Botanists," 199–202.
17. Cited from *Collected Letters of Charlotte Smith*, 332.
18. See, for example, Bailes, *Questioning Nature*, 184–85.
19. In *Conversations*, Smith more immediately attributes the use of "sylphs" to Erasmus Darwin and William Hayley.
20. Pope, *Rape of the Lock*, 1:71–72, 77–78.
21. McLaverty, *Pope, Print, and Meaning*, 38.
22. George, *Botany*, 86.
23. George, 85.
24. George, 127.
25. Smellie, *Philosophy*, 247.
26. Smellie, 251.
27. George, *Botany*, 124.
28. George, 32.
29. Pope, *Essay on Man*.
30. Schiebinger, "Gender and Natural History," 171.
31. Pascoe, "Female Botanists," 205–6.
32. For more discussion of the idea that "the narrative imagery of botany is not always conducive to poetic accuracy," see Powell, "Linnaeus, Analogy, and Taxonomy," 114.
33. Aikin, *Essay*, 25, 26.
34. Bailes, *Questioning Nature*, 94–117.
35. Knight, *Ordering the World*, 23.
36. Kelley, "Romantic Exemplarity," 238.
37. Smith, *Collected Letters*, xxv.
38. Robinson, "*River Duddon* and Wordsworth," 298, 299, 307. For a thorough discussion of similarities and differences between the poetic approaches of Smith and Wordsworth, see Labbe, *Writing Romanticism*.

39. Bailes, *Questioning Nature*, 94–117. For additional essays on Smith's treatment of birds in her poetry, see Cook, "Charlotte Smith and 'The Swallow'"; and Mellor, "Baffling Swallow."
40. Crawford, "Lyrical Strategies, Didactic Intent," 211.
41. See, for instance, "To the River Arun"; "On Leaving a Part of Sussex"; and "On Being Cautioned Against Walking on a Headland Overlooking the Sea, Because It Was Frequented by a Lunatic."
42. For example, two scholars who have investigated this subject of gender and the prospect poem include Labbe, *Romantic Visualities*; and Keith, *Poetry and the Feminine*, 80–110.
43. In addition to the imperial/economic inference, one may also be tempted to hear a slight "military strain" in these lines; while not mentioning this particular instance, as noted earlier, Kelley entertains several of the poem's allusions to the Napoleonic Wars in "Romantic Exemplarity" (238).
44. Ottum, "'Shallow' Estates," 251.
45. Smith refers to herself, or her speaker, as similarly "crush'd to the earth" in her sonnet "To the Muse," l. 9.
46. A feeling Smith expressed more directly in prefaces to her *Elegiac Sonnets*.
47. James's "law of dissociation" interestingly compares with the early nineteenth-century circular or quinary system of William MacLeay and William Swainson in which "the pattern of three major circles, the typical group, the sub-typical group, and the aberrant group (divided into three) is the pattern repeated throughout the smaller groups." While labeling its third group "aberrant" and thus sustaining primacy of traditional binaries, the quinary system (like "dissociation") nevertheless acknowledges the multiplicity of identities that are ultimately placeless in dichotomies and require the creation of new categories. Stafford, "Images of Ambiguity," 241.
48. The pronoun is used once previously in a similar "I believe" phrase regarding Yucca (l. 84n).
49. This is not the first time that Smith incorporates such seemingly hybrid or liminal organisms into her educational works. For example, in *Rambles Farther*, she writes, "I say half-animated, because testaceous fishes have little more perception than vegetables, and seem to form the link between the animal and vegetable world," and "there is a stationary, half-existing substance adhering to rocks and stones, which has been called the animal flower" (1:55).
50. Smith, *Works*, 13: 62.
51. Smith, 13:181.
52. Smith, 13:164.
53. Smith, "The Wheat-Ear," in *Poems*, l. 31n.
54. Thomson, *Before Darwin*, 225. Fascinatingly, Gosse, who sometimes intersperses short passages of poetry within his natural history texts, quotes a few of Smith's verses in his *Aquarium* (89).

55. 1 Corinthians 13:12.
56. Smith, *Natural History of Birds*, 1:4.
57. Labbe, "Transplanted into More Congenial Soil," 78.

6. Botany and Madness

1. See Whitehead, *Madness and the Romantic Poet*, 29–50.
2. Foucault discusses the affinity of the walnut and the human head and the nut's ability to cure head ailments in *The Order of Things*, 27. Also, as Burwick explains, William Blake's work addresses the "widespread cases [of ergot poisoning that] commonly followed when rainstorms felled the rye before harvest" and "contributed to mental derangement" (*Poetic Madness*, 180). Perhaps one of the most memorable botanical examples of madness in literature is William Shakespeare's Ophelia in *Hamlet* who, in her madness, gives flowers with specific meanings to Laertes, Gertrude, and Claudius, and later drowns wearing a garland or coronet of flowers.
3. Schiebinger, "Private Life of Plants," 124.
4. Whitehead, *Madness and the Romantic Poet*, 27.
5. Porter, *Madness*, 80–81; Abrams, *Mirror and the Lamp*, 170–75.
6. Kairoff explains that "Seward thus established herself through her epistolary network as a literary authority in a medium understood by contemporaries to be among the alternatives to print publication" (*Anna Seward*, 232).
7. Bailes, *Questioning Nature*, 73–93.
8. King-Hele, *Erasmus Darwin: A Life*, 107, 133.
9. Seward, *Letters*, 1:14–15. All references to this work will be cited parenthetically.
10. See, for example, Seward, *Letters*, 1:215 and 3:4–5.
11. See also Seward, 5:332–33.
12. King-Hele, *Erasmus Darwin and the Romantic Poets*, 15, 155.
13. Fletcher, *Charlotte Smith*, 327.
14. Regarding the quality of "nervousness," Kairoff writes, "Although [Charlotte] Smith's supporters praised Smith's strong and "nervous" verse, Seward denied Smith's poems those masculine—and therefore positively gendered—qualities and strove to illustrate them in her own poems" (*Anna Seward*, 174–75).
15. Burwick, *Poetic Madness*, 6.
16. King-Hele, *Erasmus Darwin and the Romantic Poets*, 13; Keats, *Letters*, 1:113.
17. Bailes, *Questioning Nature*, 81–84.
18. Mullan, "Hypochondria and Hysteria," 148. On "nerve theory," see also Rousseau, "Nerves, Spirits and Fibres"; Ahern, *Affected Sensibilities*, 16–17; and Crouch, "Nerve Theory and Sensibility," 212.

19. Labbe, "Pathological Sensibility," 356. Wilson also discusses the nerves in relation to sensibility and what she calls "sensibility's pathological turn" in the early nineteenth century; I return to this topic in my conclusion ("End of Sensibility," 276).
20. Quoted in Jackson, *Science and Sensation*, 146.
21. See Jackson, 146.
22. For more on Seward as a poet of sensibility, see, for instance, Kairoff, *Anna Seward*, 226.
23. King-Hele, "Erasmus Darwin, Man of Ideas," 153. See also Uglow, *Lunar Men*. For a recent study of instances of "insensibility" in the eighteenth-century novel, see Lee, *Failures of Feeling*.
24. Darwin, *Life of Erasmus Darwin*, 75–76; Barnard, *Anna Seward*, 141–42.
25. King-Hele, *Erasmus Darwin: A Life*, 327.
26. King-Hele, 327.
27. King-Hele, *Erasmus Darwin and the Romantic Poets*, 92–93; King-Hele, *Erasmus Darwin: A Life*, 287.
28. Logan, *Poems on Several Occasions*, 21.
29. Seward, *Original Sonnets*, 73.
30. See Keith's sensitive reading of this poem in "Poetry, Sentiment, and Sensibility," 138. In line with a suggestion made by Keith, Freeman also writes that Seward "likens the poppy flower to the female sex organ" ("Opium Use," 13).
31. Ingram, with Faubert, *Cultural Constructions of Madness*, 150.
32. Cowper, *Task*, 29–30.
33. King-Hele, *Erasmus Darwin and the Romantic Poets*, 149.
34. King-Hele, 40.
35. Haut, "Reading Flora," 243.
36. Myrone, "Gothic Romance and the Quixotic Hero."
37. Myrone.
38. Myrone, *Henry Fuseli*, 47.
39. Trott, "Wordsworth's *Loves of the Plants*," 152.
40. Seward writes that such Calvinistic mysticism appeals to "the same disposition which makes children delight . . . in perceiving objects of terror presented to their imagination . . . but no mischievous or obstinate child is rendered gentle or docile by the dread of specters; neither have the fanatic tenets any tendency to reclaim from vice or irreligious thoughtlessness" (*Letters*, 5:246).
41. See also Wood, "Female Penseroso." Although various scholars reference Seward's sociable poetry, it is also useful to note Kairoff's discussion of solitude in "Samuel Johnson and Anna Seward."
42. See *OED* entries for "daffodil" (https://www.oed.com/view/Entry/46852?redirectedFrom=daffodil#eid); "daffodilly" (https://www.oed.com/view

/Entry/46853); "daffy" (https://www.oed.com/view/Entry/46856); and "daft" (https://www.oed.com/view/Entry/46858), accessed 4 September 2022.
43. Hazlitt, *Lectures,* 323; emphasis mine. Quoted in Ross, *Contours of Masculine Desire,* 255.
44. Jackson, *Science and Sensation,* 88.
45. Coleridge, *Biographia Literaria,* 315.
46. Levinson, "Motion and a Spirit"; Levinson, "Of Being Numerous"; Ottum and Reno, "Introduction," 1.
47. King-Hele, *Erasmus Darwin and the Romantic Poets,* 64; Wordsworth, "Lines Written in Early Spring," ll. 11–12.
48. Blick, "Flashing Flowers." King-Hele suggests that William Blake, too, had this phenomenon in mind and drew on Darwin's notes about it in one section of *Visions of the Daughters of Albion* (1793); see *Erasmus Darwin and the Romantic Poets,* 44–45.
49. Blick, 110–11, 113.
50. Blick, 112.
51. Mahood, *Poet as Botanist,* 40.
52. Burwick, *Poetic Madness,* 14.
53. Pinch, *Strange Fits of Passion,* 73.
54. Pinch, 73; Hartman, *Wordsworth's Poetry,* 3, 6. Jackson instead argues that Wordsworth's egotism, inwardness, or interiority could be understood as "always-already social" (*Science and Sensation,* 7).
55. Keats to Woodhouse, 27 October 1818; Keats to Reynolds, 3 February 1818.
56. Whitehead, *Madness and the Romantic Poet,* 183.
57. Bate, *John Clare,* 187, 119, 405, 380. In this chapter, this source will be cited parenthetically as "Bate."
58. *Edinburgh Magazine* 2 (April 1818): 316; Storey, "Clare and the Critics," 36.
59. Haughton, "Progress and Rhyme," 53.
60. King-Hele, *Erasmus Darwin and the Romantic Poets,* 265–66.
61. J. W. and A. Tibble, eds., *Prose of John Clare,* 51. Clare acquired James Smith's *Compendium of the English Flora* (1829) and noted the titles of other botanical texts, "but there is no evidence that he read these" texts (Mahood, *Poet as Botanist,* 129).
62. J. W. and A. Tibble, 53–54.
63. Robinson and Powell, eds., *John Clare by Himself,* 189.
64. Grainger, ed., *Natural History Prose Writings,* 38–39.
65. Clare, *Letters,* 403. Clare possessed copies of both Darwin's *Botanic Garden* and *Temple of Nature.*
66. Mahood, *Poet as Botanist,* 112.
67. Mahood, 145; Clare, "To the Glow-worm," l. 3.

68. When Clare was admitted to Matthew Allen's private asylum in Epping Forest in 1837, his doctors signing the certificate of insanity answered that the cause of his illness was "hereditary," and, as mentioned earlier, Darwin considered this a prime indicator for potential madness (Bate, 409). Darwin also classified as insanity an excess of *nostalgia*, quoting verses from Goldsmith and defining the term as an "unconquerable desire of returning to one's native country," and numerous scholars have applied this sentiment to Clare's descriptions of, and later longings for, his native Helpston (*Zoonomia*, 2:367). Additionally, Darwin categorized as madness the fear of being "infected with the venereal disease, when they have only deserved it," and this also could be applied to Clare, as he believed that his experiences of madness may be due to a sexually transmitted disease (2:382; Bate, 411). Since Clare's doctors viewed the actual effects of, and perpetual efforts to prevent, his and his family's poverty as a major cause of his illness, Darwin's category of the "fear of poverty" also may be applied (2:376; Bate, 252). And, again, Darwin even classified some kinds of poetry as insanity, in the form of "sentimental love" as "described in its excess by romance-writers, and poets" in which "the object of love is beauty, and as our perception of beauty consists in a recognition by the sense of vision of those objects, which have before inspired our love, by the pleasure they have afforded to many of our senses" (2:363). Through this definition and the way in which Clare describes his feelings for nature, it seems as applicable to his love and versification of natural objects. See also Roy Porter, "All Madness for Writing," 264.
69. Whitehead, *Madness and the Romantic Poet*, 92.
70. Burwick, *Poetic Madness*, 265, 266.
71. Clare, letter to "My Dear Wife Mary," 1841; quoted in Burwick, *Poetic Madness*, 269.
72. Gaull, "Clare and 'the Dark System,'" 291.
73. See Keith, "Poetry, Sentiment, and Sensibility," 137.
74. Barnard, *Anna*, 145.
75. Whitehead, *Madness and the Romantic Poet*, 2–3.
76. Whitehead, 126–27.
77. Porter, *Madness*, 81.
78. "Walking Stewart," 258.
79. For example, nineteenth-century scientists, specifically geologists, accused contemporary poets of madness, see Bailes, *Questioning Nature*, 157.

Conclusion

1. Ross, "Twin Labourers," 34–35. Ross particularly examines the competition between Humphry Davy and William Wordsworth. It is also worth noting that endeavors toward professionalization within the sciences

coincided with literary authors' public discussions of professional income and the introduction of domestic and international copyright bills in Parliament, especially from 1837 onward. Indeed, Wordsworth, Southey, and Carlyle each publicly urged perpetual copyright laws at this time; see Peterson, *Becoming a Woman of Letters*, 35; and Bailes, *Questioning Nature*, 198.

2. Macfarlane, *Original Copy*; Bailes, *Questioning Nature*; for other studies of the intertextualities of male Romantics, see Mazzeo, *Plagiarism and Literary Property*; Leader, *Revision and Romantic Authorship*; and Stillinger, *Multiple Authorship*.
3. See Murphy, *In Science's Shadow*, 1; and Battersby, *Gender and Genius*. See also the various nineteenth-century male scientists quoted in Bailes, *Questioning Nature*, 196.
4. Mullan, "Hypochondria and Hysteria," 148; Showalter, *Female Malady*.
5. Brown, "Hamilton and Wordsworth," 479; Bailes, *Questioning Nature*, 181.
6. Fulford, "De Quincey's Literature of Power," 158. See also Bate, "Literature of Power."
7. Hazlitt, *Lectures*, 323.
8. Hazlitt, 309.
9. Hazlitt, 323.
10. Hazlitt, 286.
11. Hazlitt, 287–88.
12. Hazlitt, 290–92.
13. Hazlitt, "Of Great and Little Things" (1821), in *Complete Works*, 8:226–42, 236.
14. Bate discusses Hazlitt's influence on conceptions of "power" as well as De Quincey's interactions with Coleridge's writings ("Literature of Power").
15. See, for example, Darwin, *Zoonomia*, 1:354; and chapter 6 of this book.
16. Brown, "Hamilton and Wordsworth," 481.
17. De Quincey originally published "False Distinctions" in June 1824; I am quoting from its reprinting in De Quincey, *Suspiria De Profundis*, 503.
18. Kitson, *Romantic Literature*, 205. See also Kitson, "Strange Case."
19. Kitson, *Romantic Literature*, 207.
20. De Quincey, *Suspiria De Profundis*, 506–7; for a slightly earlier example of this argument regarding women's imagination, see Duff, *Letters*, 29.
21. North, "De Quincey and the Inferiority of Women," 331.
22. Mullan, "Hypochondria and Hysteria," 159.
23. In contrast, in 1810, Anna Barbauld had published an essay concluding "that the best practitioners of the art of the novel in her day . . . were mainly women," and implied "that female novelists [had] become the political as well as cultural arbiters of the British nation" (Mellor, "Were Women Writers 'Romantics'?," 399, 400). Barbauld, "Essay," 56, 59.

24. Bailes, *Questioning Nature*, 202.
25. Wordsworth, 23 December 1829, *Letters*, 2:185. This and the subsequent paragraph benefit from Behrendt, "William Wordsworth and Women Poets," 635–50.
26. Wordsworth, Autumn 1841, 4:245.
27. Kennedy, "Wordsworth, Hemans," 268.
28. Hemans, *Felicia Hemans*, 68.
29. Godwin, *Poetical Works*, vii.
30. Wordsworth, 15 December 1837, *Letters*, 3:491.
31. Mitford, 13 September 1817, *Life*, 1:270.
32. Robinson, *Correspondence*, 1:53.
33. Wordsworth, *Letters*, 12 January 1829, 2:4–5; see, for example, Pinch, *Strange Fits of Passion*, 74; and Behrendt, "William Wordsworth and Women Poets," 635–50.
34. Wordsworth, 16 October 1829, *Letters*, 2:157.
35. Behrendt, "William Wordsworth and Women Poets," 637, 639, 646.
36. Wordsworth, 10 May 1830, *Letters*, 2:260.
37. Wordsworth, 2:259.
38. Wordsworth, 19 April 1830, 2:235.
39. Behrendt, "William Wordsworth and Women Poets," 635.
40. Quoted in Burwick, *Poetic Madness*, 256.
41. John Ruskin, "Of Queens' Gardens," 52.
42. McGrath, "Thomas De Quincey," 860.
43. De Quincey, "Works of Alexander Pope," 16:336. All references to this work will be cited parenthetically.
44. Jackson, *Science and Sensation*, 11.
45. On women's prominence as authors of cookbooks in the late eighteenth and early nineteenth centuries, see, for example, Merrett, "Culinary Art."
46. North, "De Quincey and the Inferiority of Women," 329.
47. Barrell, *Infection of Thomas De Quincey*, 163.
48. Fulford, "De Quincey's Literature of Power," 159, 163; Whale, "De Quincey and Men (of Letters)," 94–95.
49. Murphy, *In Science's Shadow*, 1; see also, for example, Battersby, *Gender and Genius*.
50. Darwin, *Descent of Man*, 326–27; for similar statements from male scientific thinkers of the era, see also, for example, Showalter, *Female Malady*, 122–23; and Bailes, *Questioning Nature*, 196.
51. Kitson, *Romantic Literature*, 16, 17.
52. Mellor asserts that "writers like Wordsworth, Scott, Burke, De Quincey, and James Mill promoted a concept of an autonomous self that defined itself in opposition, even in hostile antagonism, to the cultural or ethnic other," while "women writers between 1700 and 1900 displayed far more

interest in and sympathy for racial and cultural difference" ("Were Women Writers 'Romantics'?," 402–3). See also Makdisi, *Romantic Imperialism*, 120–21; and Chander, *Brown Romantics*, 2.
53. Looser, "Why I'm Still Writing," 224.
54. See, for example, Huxley regarding the erasure of this era's scientific literature more generally (*Life and Letters*, 2:337; quoted in Bailes, *Questioning Nature*, 202). Earlier in my current study, I also have quoted Hazlitt, Clare, John Aikin, and Charles Darwin on the relative obscurity of Erasmus Darwin's poetry in particular, as early as 1815.

BIBLIOGRAPHY

Primary Sources

Addison, Joseph. *Spectator*, no. 69 (19 May 1711).
Aikin, John. *The Calendar of Nature; Designed for the Instruction and Entertainment of Young Persons*. Warrington: W. Eyres, 1784.
———. *Essay on the Application of Natural History to Poetry*. Warrington: J. Johnson, 1777.
———. *An Essay on the Plan and Character of Thomson's Seasons*. London: J. Murray, 1788.
———. *General Biography; or Lives, Critical and Historical*. 10 vols. London: G. G. & J. Robinson, 1799–1815.
Analytical Review 15 (1793): 287–93.
Anti-Jacobin Review and Magazine; or, Monthly Political and Literary Censor. January–April 1800.
Barbauld, Anna. *Lessons for Children from Three to Four Years Old*. London: J. Johnson, 1778–79.
———. *Pleasures of Imagination by Mark Akenside, A Critical Essay on the Poem*. London, 1795.
———. *The Poems of Anna Letitia Barbauld*. Edited by William McCarthy and Elizabeth Kraft. Athens: University of Georgia Press, 1994.
British Critic 20 (1802).
British Critic 30 (1807).
Browne, Patrick. *The Civil and Natural History of Jamaica*. London: B. White, [1756] 1789.
Buffon, Georges-Louis Leclerc, comte de. *Natural History, General and Particular*. 9 vols. Edited by William Smellie. Edinburgh: W. Creech, 1780–85.
Burn, Andrew. *A Second Address to the People of Great Britain: Containing a New and Most Powerful Argument to Abstain from the Use of West India Sugar*. London: M. Gurney, 1792.
Burnet, Thomas. *The Sacred Theory of the Earth: Containing an Account of the Original of the Earth, and of All the General Changes*. 2 vols. London: John Hooke, [1684] 1719.
Clare, John. *The Letters of John Clare*. Edited by Mark Storey. Oxford: Clarendon, 1985.
———. *The Prose of John Clare*. Edited by J. W. and Anne Tibble. London: Routledge & Kegan Paul, [1951] 1970.

———. *The Rural Muse*. London: Whittaker, 1835.

———. *The Shepherd's Calendar*. London: J. Taylor, 1827.

Coleridge, Samuel Taylor. *Biographia Literaria*. Edited by Adam Roberts. Edinburgh: Edinburgh University Press, [1817] 2014.

———. *Collected Letters of S. T. Coleridge*. Edited by F. L. Griggs. 6 vols. Oxford: Oxford University Press, 1956–71.

———. *Lectures, 1808–1819: On Literature*. 2 vols. Edited by R. A. Foakes. Vol. 54 of *The Collected Works of Samuel Taylor Coleridge*. Princeton, NJ: Princeton University Press, 1987.

———. *Letters of Samuel Taylor Coleridge*. 2 vols. Edited by Ernest Hartley Coleridge. Boston, 1895.

———. *The Notebooks of Samuel Taylor Coleridge, Volume 2: 1804–1808 Text*. Edited by Kathleen Coburn. New York: Routledge, [1962] 2002.

Cowper, William. *The Task, A Poem, in Six Books*. London: J. Johnson, 1785.

Cox, James. *Descriptive Catalogue of the Several Superb and Magnificent Pieces of Mechanism and Jewellery, Exhibited in the Museum, at Spring-Gardens, Charing-Cross*. London: n.p., 1772.

Darwin, Charles. *The Descent of Man, and Selection in Relation to Sex*. Princeton, NJ: Princeton University Press, [1871] 1981.

———. *The Life of Erasmus Darwin*. Edited by Desmond King-Hele. Cambridge: Cambridge University Press, 2003.

Darwin, Erasmus. *The Botanic Garden, a Poem in Two Parts. Part I. Containing the Economy of Vegetation. Part II. The Loves of the Plants. With Philosophical Notes*. London: J. Johnson, 1791.

———. *The Botanic Garden; Part II. Containing the Loves of the Plants, a Poem. With Philosophical Notes*. London: J. Johnson, 1789.

———. *The Letters of Erasmus Darwin*. Edited by Desmond King-Hele. Cambridge: Cambridge University Press, 1981.

———. *Phytologia; or The Philosophy of Agriculture and Gardening*. Dublin: P. Byrne, 1800.

———. *The Temple of Nature; or, The Origin of Society: A Poem, with Philosophical Notes*. London: J. Johnson, 1803.

———. *Zoonomia; or The Laws of Organic Life*. 2 vols. London: J. Johnson, 1794–96.

De Quincey, Thomas. "Letters to a Young Man Whose Education Has Been Neglected." In *The Works of Thomas De Quincey*, vol. 3, edited by Frederick Burwick, 39–97. London: Pickering & Chatto, 2000.

———. *Suspiria De Profundis: Being a Sequel to the Confessions of an English Opium-Eater*. Edinburgh, 1871.

———. "The Works of Alexander Pope, Esquire." In *The Works of Thomas De Quincey*, vol. 16, 332–64. London: Pickering & Chatto, 2003.

Duff, William. *Letters on the Intellectual and Moral Character of Women*. Aberdeen: J. Chalmers, 1807.

Edinburgh Magazine 2 (April 1818).

Godwin, Catherine Grace. *The Poetical Works of the Late Catherine Grace Godwin, with a Sketch of Her Life*. Edited by A. Cleveland Wigan. London: n.p., 1854.

Goethe, Johann Wolfgang von. *Versuch die Metamorphose der Pflanzen zu erklaren* [*Metamorphosis of Plants*]. Gotha, 1790.

Goldsmith, Oliver. *A History of the Earth, and Animated Nature*. London: J. Nourse, 1774.

Grainger, James. *The Sugar-Cane: A Poem, in Four Books*. London: R. & J. Dodsley, 1764.

Gray, Samuel Frederick. *Natural Arrangement of British Plants, in Two Volumes*. London: Baldwin, 1821.

Green, Thomas. *The Universal Herbal; or, Botanical, Medical, and Agricultural Dictionary*. 2 vols. Liverpool: Caxton, 1820.

Hazlitt, William. *The Complete Works*. Edited by P. P. Howe. London: J. M. Dent & Sons, 1931.

———. *Lectures on the English Poets*. London: Taylor & Hessey, 1818.

Hemans, Felicia. *Felicia Hemans: Selected Poems, Letters, Reception Materials*. Edited by Susan J. Wolfson. Princeton, NJ: Princeton University Press, 2000.

Huxley, Thomas H. *Life and Letters of Thomas Huxley*. 2 vols. Edited by Leonard Huxley. Macmillan, 1990.

Johnson, Samuel. *Idler*, no. 97 (23 February 1760).

Keats, John. *The Letters of John Keats*. 2 vols. Edited by H. E. Rollins. Cambridge, MA: Harvard University Press, 1958.

Kerr, Robert. *Memoirs of the Life, Writings, and Correspondence of William Smellie*. Edinburgh: John Anderson, 1811.

Lamb, Charles. *The Letters of Charles and Mary Anne Lamb*. Edited by Edwin W. Marrs Jr. Ithaca, NY: Cornell University Press, 1976.

Linnaeus, Carl. *The Elements of Botany*. Edited by Hugh Rose. London: T. Cadell, 1775.

———. *Philosophia Botanica*. Edited by Stephen Freer. Oxford: Oxford University Press, 2005.

Lobb, Richard. *The Contemplative Philosopher*. 2 vols. London: Sherwood, 1800.

Locke, John. *An Essay Concerning Human Understanding*. Edited by Peter H. Nidditch. Oxford: Clarendon, [1690] 1975.

Logan, Maria. *Poems on Several Occasions*. York: Wilson, 1793.

Long, Edward. *The History of Jamaica, or General Survey of the Antient and Modern State of That Island with Reflections on Its Situation, Settlements,*

Inhabitants, Climate, Products, Commerce, Laws, and Government. London: T. Lowndes, 1774.

Mathias, Thomas James. *The Pursuits of Literature, or What You Will: A Satirical Poem in Dialogue.* London: T. Becket, 1794.

Merian, Maria Sibylla. *Metamorphosis Insectorum Surinamensium.* Amsterdam: Dritte Auflage, [1705] 1994.

Milne, Colin. *A Botanical Dictionary: Or, Elements of Systematic and Philosophical Botany.* London: William Griffin, 1770.

Mitford, Mary Russell. *The Life of Mary Russell Mitford, Told by Herself in Letters to Her Friends.* 3 vols. Edited by Rev. A. G. L'Estrange. London: Harper, 1870.

Moir, David Macbeth. *Sketches of the Poetical Literature of the Past Half-Century.* Edinburgh: W. Blackwood, 1851.

Montagu, Mary Wortley. *Letters.* 3 vols. London: T. Becket, 1763.

Montesquieu, Charles-Louis de Secondat, baron de. *The Spirit of the Laws.* London: n.p., 1750.

Monthly Repository and Review of Theology and General Literature 1 (1827).

Monthly Review 56 (May–August 1808).

Moryson, Fynes. *An History of Ireland, from the Year 1599 to 1603.* 2 vols. Dublin: n.p., 1735.

Mundy, Francis Nöel Clarke. *Needwood Forest.* Lichfield: John Jackson, 1776.

———. *Needwood Forest and the Fall of Needwood, with Other Poems.* Derby: T. Richardson, 1830.

Owenson, Sydney, Lady Morgan. *The Lay of an Irish Harp; or, Metrical Fragments.* London: Richard Phillips, 1807.

———. Review of J. G. C. Feuillide's *Ireland. Athenaeum,* no. 598 (13 April 1839): 273–74.

———. *The Wild Irish Girl: A National Tale.* Edited with introduction and notes by Kathryn Kirkpatrick. Oxford: Oxford University Press, 1999.

Paley, William. *Natural Theology, or Evidences of the Existence and Attributes of the Deity.* London: R. Faulder, 1802.

Penn, John. *Poetical Miscellanies. Including Translations from Petrarch.* London: n.p., 1797.

Pennant, Thomas. *Arctic Zoology.* 2 vols. London: Henry Hughs, 1784–85.

———. *A Tour in Scotland, and Voyage to the Hebrides.* Chester: John Monk, 1774.

———. *The View of Hindoostan.* 2 vols. London: Henry Hughs, 1798.

Polwhele, Richard. *Unsex'd Females: A Poem, Addressed to the Author of The Pursuits of Literature.* London: Cadell & Davies, 1798.

Pope, Alexander. *Essay on Man. Addressed to a Friend.* London: J. Wilford, 1733–34.

———. *The Rape of the Lock. An Heroical Poem in Five Cantos.* London: Bernard Lintott, 1714.

Rawdon, Elizabeth, Countess of Moira. "Particulars Relative to a Human Skeleton." *Archaeologia* 7 (1785).
Riddell, Maria. *Voyages to the Madeira, and Leeward Caribbean Isles: with Sketches of the Natural History of these Islands*. Edinburgh: Peter Hill, 1792, 1802.
Riddell, Maria, ed. *The Metrical Miscellany: Consisting Chiefly of Poems Hitherto Unpublished*. London: T. Cadell & W. Davies, 1802.
Robinson, Henry Crabb. *The Correspondence of Henry Crabb Robinson with the Wordsworth Circle (1808–1866)*. 2 vols. Oxford: Clarendon, 1927.
Rousseau, Jean-Jacques. *Letter on the Elements of Botany, Addressed to a Lady*. Edited by Thomas Martyn. London: B. White, 1785.
Ruskin, John. "Of Queens' Gardens." In *Sesame and Lilies*. n.p., [1864] 1970.
Saint-Pierre, Jacques-Henri Bernardin de. *A Voyage to the Island of Mauritius*. Edited by John Parish. London: W. Griffin, 1775.
Schaw, Janet. *Journal of a Lady of Quality: Being the Narrative of a Journey from Scotland to the West Indies, North Carolina, and Portugal, in the Years 1774 to 1776*. Edited by Evangeline Walker Andrews and Charles McLean Andrews. New Haven, CT: Yale University Press, [1921] 1934.
Seward, Anna. *Letters of Anna Seward: Written between the Years 1784 and 1807*. 6 vols. Edinburgh: A. Constable, 1811.
———. *Memoirs of the Life of Dr. Darwin, Chiefly during his Residence in Lichfield*. London: J. Johnson, 1804.
———. *Original Sonnets on Various Subjects; and Odes Paraphrased from Horace*. London: G. Sael, 1799.
Shoberl, Frederic. *The Language of Flowers*. Revised by Louise Cortambert. London: Saunders & Otley, 1835.
Smellie, William. *Philosophy of Natural History*. Edinburgh: Printed for C. Elliot & T. Kay; T. Cadel; G. G. J. & J. Robinsons, 1790.
Smith, Charlotte. *The Collected Letters of Charlotte Smith*. Edited by Judith Phillips Stanton. Bloomington: Indiana University Press, 2003.
———. *A Natural History of Birds, Intended Chiefly for Young Persons*. 2 vols. London: n.p., 1807, 1815.
———. *The Poems of Charlotte Smith*. Edited by Stuart Curran. Oxford: Oxford University Press, 1993.
———. *Rambles Farther: A Continuation of Rural Walks: In Dialogues. Intended for the Use of Young Persons*. 2 vols. London: Cadell & Davies, 1796.
———. *The Works of Charlotte Smith*. 14 vols. Edited by Stuart Curran. London: Pickering & Chatto, 2007.
Southey, Robert. "Collected Works of the Late Dr. Sayers." *Quarterly Review* 12 (1814): 60–90.
Spenser, Edmund. *A View of the Present State of Ireland*. In *The Works of Spenser*. 6 vols. London: J. & R. Tonson/S. Draper, [1596] 1750.
Stock, J. E. *Memoirs of the Life of Thomas Beddoes*. London: John Murray, 1811.

Swift, Jonathan. *A Modest Proposal.* Dublin: S. Harding, 1729.
Taylor, Joseph. *The Complete Weather Guide.* London: Printed for John Harding, 1812.
Thomson, James. *The Seasons and The Castle of Indolence.* Edited by James Sambrook. Oxford: Clarendon, 1991.
Thornton, Robert John. *New Illustration of the Sexual System of Carolus von Linnaeus.* London: T. Bensley, 1799–1806.
"Walking Stewart." *London Magazine* 8 (1823): 253–60.
Walpole, Horace. *The Letters of Horace Walpole.* Edited by Helen Toynbee and Paget Toynbee. Oxford: Clarendon, 1905.
White, Gilbert. *The Natural History of Selborne.* Edited by Paul Foster. Oxford: Oxford University Press, [1789] 1993.
Withering, William. *A Botanical Arrangement of All the Vegetables Naturally Growing in Great Britain.* 2 vols. Birmingham: M. Swinney, 1776.
Wollstonecraft, Mary. *Letters Written in Sweden, Norway, and Denmark.* London: J. Johnson, 1796.
Wordsworth, William. *The Letters of William and Dorothy Wordsworth: The Later Years.* 4 vols. Edited by Alan G. Hill. Oxford: Clarendon, 1978–88.
———. *Lyrical Ballads, with Other Poems.* 2 vols. London: T. N. Longman & O. Rees, 1800.
———. *Poems, Chiefly of Early and Late Years.* London: Edward Moxon, 1842.
———. *Poetical Works.* Edited by Ernest de Selincourt and Helen Darbishire. 5 vols. Oxford: Clarendon, 1940–49.
———. *Poetical Works.* Edited by Ernest de Selincourt and Helen Darbishire. Oxford: Clarendon, 1952–59.
———. *William Wordsworth: The Major Works.* Edited by Stephen Gill. Oxford: Oxford University Press, [1984] 2011.

Secondary Sources

Abrams, M. H. *The Mirror and the Lamp: Romantic Theory and the Critical Tradition.* Oxford: Oxford University Press, [1953] 1971.
Ahern, Stephen. *Affected Sensibilities: Romantic Excess and the Genealogy of the Novel, 1680–1810.* New York: AMS, 2007.
Andrews, Kerri. "Charlotte Smith." In *The Encyclopedia of British Literature: 1660–1789*, vol. 3, edited by Gary Day and Jack Lynch, n.p. Malden, MA: Wiley Blackwell, 2015.
Avramescu, Catalin. *An Intellectual History of Cannibalism.* Edited by Alistair Ian Blyth. Princeton, NJ: Princeton University Press, 2009.
Bailes, Melissa. "Cultivated for Consumption: Botany, Colonial Cannibalism, and National/Natural History in Sydney Owenson's *The Wild Irish Girl.*" *Eighteenth Century: Theory and Interpretation* 59, no. 4 (Winter 2018): 513–33.

———. "Hybrid Britons: West Indian Colonial Identity and Maria Riddell's Natural History." *European Romantic Review* 20, no. 2 (2009): 207–17.

———. "Linnaeus's Botanical Clocks: Chronobiological Mechanisms in the Scientific Poetry of Erasmus Darwin, Charlotte Smith, and Felicia Hemans." *Studies in Romanticism* 56, no. 2 (Summer 2017): 223–52.

———. "Literary Plagiarism and Scientific Originality in the 'Trans-Atlantic Wilderness' of Goldsmith, Aikin, and Barbauld." *Eighteenth-Century Studies* 49, no. 2 (2016): 265–79.

———. "'On the green margin': Science, Gender, and Originality in Charlotte Smith's 'Flora.'" In *Placing Charlotte Smith*, edited by Elizabeth A. Dolan and Jacqueline M. Labbe, 159–80. Bethlehem, PA: Lehigh University Press, 2020.

———. *Questioning Nature: British Women's Scientific Writing and Literary Originality, 1750–1830*. Charlottesville: University of Virginia Press, 2017.

———. "Transformations of Gender and Race in Maria Riddell's Transatlantic Biopolitics." *Eighteenth-Century Fiction* 32, no. 1 (Fall 2019): 123–44.

Bakhtin, Mikhail. *The Dialogic Imagination: Four Essays by M. M. Bakhtin*. Edited by Michael Holquist. Translated by Caryl Emerson and Michael Holquist. Austin: University of Texas Press, 1994.

Barker-Benfield, G. J. *The Culture of Sensibility: Sex and Society in Eighteenth-Century Britain*. Chicago: University of Chicago Press, 1992.

Barnard, Teresa. *Anna Seward: A Constructed Life; A Critical Biography*. Farnham: Ashgate, 2009.

———. "Anna Seward's Hidden Words: Female Interventions into Male Writing." *Women's Writing* 19, no. 4 (2012): 417–33.

Barratt, Glynn R. "James Thomson in Russia: The Changing of *The Seasons*." *Comparative Literature Studies* 12, no. 4 (December 1975): 367–73.

Barrell, John. *The Infection of Thomas de Quincey*. New Haven, CT: Yale University Press, 1991.

———. *Poetry, Language, and Politics*. Manchester: Manchester University Press, 1988.

Bate, Jonathan. *John Clare: A Biography*. New York: Farrar, Straus & Giroux, 2003.

———. "The Literature of Power: Coleridge and De Quincey." In *Coleridge's Visionary Languages*, edited by Tim Fulford and Morton D. Paley, 137–50. Woodbridge: D. S. Brewer, 1993.

Battersby, Christine. *Gender and Genius: Toward a Feminist Aesthetic*. London: Women's Press, 1989.

Behrendt, Stephen C. "William Wordsworth and Women Poets." *European Romantic Review* 23, no. 6 (December 2012): 635–50.

Belanger, Jacqueline. *Critical Receptions: Sydney Owenson, Lady Morgan*. Bethesda, MD: Academica, 2007.

Bergstrom, Carson. "James Thomson's *A Poem Sacred to the Memory of Sir Isaac Newton* and the Revisions to *The Seasons*: New Science and Poetics in the Eighteenth Century." In *Experiments in Genre in Eighteenth-Century Literature*, edited by Sandro Jung, 33–60. Lebanon, NH: Academia, 2011.

Bewell, Alan. "Erasmus Darwin's Cosmopolitan Nature." *ELH* 76, no. 1 (Spring 2009): 19–48.

———. "Hyena Trouble." *Studies in Romanticism* 53, no. 3 (2014): 369–97.

———. "'Jacobin Plants': Botany as Social Theory in the 1790s." *Wordsworth Circle* 20, no. 3 (1989): 132–39.

———. "Jefferson's Thermometer: Colonial Biogeographical Constructions of the Climate of America." In *Romantic Science: The Literary Forms of Natural History*, edited by Noah Heringman, 111–38. Albany: State University of New York Press, 2003.

———. *Natures in Translation: Romanticism and Colonial Natural History*. Baltimore, MD: Johns Hopkins University Press, 2017.

Bhabha, Homi K. *The Location of Culture*. London: Routledge, [1994] 2012.

———. "The Postcolonial Critic." *Arena* 96 (1991): 47–63.

Blick, Fred. "Flashing Flowers and Wordsworth's 'Daffodils.'" *Wordsworth Circle* 48, no. 2 (Spring 2017): 110–15.

Bohls, Elizabeth A. *Slavery and the Politics of Place: Representing the Colonial Caribbean, 1770–1833*. Cambridge: Cambridge University Press, 2014.

Bowden, Betsy, ed. *Eighteenth-Century Modernizations from the Canterbury Tales*. Rochester, NY: Boydell & Brewer, 1991.

Bowler, Peter J. *Evolution: The History of an Idea*. 3rd ed. Berkeley: University of California Press, 2003.

Braun, Heather. "The Seductive Masquerade of *The Wild Irish Girl*: Disguising Political Fear in Sydney Owenson's National Tale." *Irish Studies Review* 13, no. 1 (2005): 33–43.

Braunschneider, Theresa. "The Lady and the Lapdog: Mixed Ethnicity in Constantinople, Fashionable Pets in Britain." In *Humans and Other Animals in Eighteenth-Century British Culture: Representation, Hybridity, and Ethics*, edited by Frank Palmeri, 31–48. Burlington, VT: Ashgate, 2006.

Brown, Creighton Nicholas. "The Hunger: The Power and Politics of a (Post) Colonial Cannibal." In *Diasporic Identities and Empire: Cultural Contentions and Literary Landscapes*, edited by Anastasia Nicéphore and David Brooks, 53–73. Newcastle upon Tyne: Cambridge Scholars, 2013.

Brown, Daniel. "William Rowan Hamilton and William Wordsworth: The Poetry of Science." *Studies in Romanticism* 51 (Winter 2012): 475–501.

Browne, Janet. "Botany for Gentlemen: Erasmus Darwin and *The Loves of the Plants*." *Isis* 80, no. 4 (1989): 593–621.

Burwick, Frederick. *Poetic Madness and the Romantic Imagination*. University Park: Penn State University Press, 1996.

Campbell, Mary. *Lady Morgan: The Life and Times of Sydney Owenson*. London: Pandora, 1988.
Casid, Jill H. *Sowing Empire: Landscape and Colonization*. Minneapolis: University of Minnesota Press, 2005.
Chakrabarty, Dipesh. *Provincializing Europe: Postcolonial Thought and Historical Difference*. Princeton, NJ: Princeton University Press, 2000.
Chander, Manu Samriti. *Brown Romantics: Poetry and Nationalism in the Global Nineteenth Century*. Lewisburg, PA: Bucknell University Press, 2017.
Chandler, James. "The Question of Sensibility." *New Literary History* 49 (2018): 467–92.
Chico, Tita. *The Experimental Imagination: Literary Knowledge and Science in the British Enlightenment*. Stanford, CA: Stanford University Press, 2018.
Chiles, Katy L. *Transformable Race: Surprising Metamorphoses in the Literature of Early America*. Oxford: Oxford University Press, 2014.
Clarke, Norma. *Ambitious Heights: Writing, Friendship, Love—The Jewsbury Sisters, Felicia Hemans, and Jane Welsh Carlyle*. London: Routledge, 1990.
Cloudsley-Thompson, J. L. *Biological Clocks: Their Functions in Nature*. London: Weidenfeld & Nicholson, 1980.
Cohen, Ralph. *The Art of Discrimination: Thomson's* The Seasons *and the Language of Criticism*. Oakland: University of California Press, 1964.
Cole, Lucinda, and Robert Markley. "Human, Animal, and Machine in the Seventeenth Century." In *A Companion to British Literature, Volume II: Early Modern Literature, 1450–1660*, edited by Robert DeMaria Jr., Heesock Chang, and Samantha Zacher, 375–90. Oxford: John Wiley & Sons, 2014.
Coleman, Deirdre. "Janet Schaw and the Complexions of Empire." *Eighteenth-Century Studies* 36, no. 2 (2003): 169–93.
Colley, Linda. *Britons: Forging the Nation, 1707–1837*. 2nd ed. New Haven, CT: Yale University Press, 1992.
Connell, Philip. "Newtonian Physico-Theology and the Varieties of Whiggism in James Thomson's *The Seasons*." *Huntington Library Quarterly* 72, no. 1 (2009): 1–28.
Cook, Elizabeth Heckendorn. "Charlotte Smith and 'The Swallow': Migration and Romantic Authorship." *Huntington Library Quarterly* 72, no. 1 (2009): 48–67.
———. "'Perfect' Flowers, Monstrous Women: Eighteenth-Century Botany and the Modern Gendered Subject." In *"Defects": Engendering the Modern Body*, edited by Helen Deutsch and Felicity Nussbaum, 252–79. Ann Arbor: University of Michigan Press, 2000.
Crawford, Rachel. "Lyrical Strategies, Didactic Intent: Reading the Kitchen Garden Manual." In *Romantic Science: The Literary Forms of Natural History*, edited by Noah Heringman. 199–222. Albany: State University of New York Press, 2003.

———. *Poetry, Enclosure, and the Vernacular Landscape, 1700–1830*. Cambridge: Cambridge University Press, 2002.

Crouch, Eleanor C. L. "Nerve Theory and Sensibility: 'Delicacy' in the Work of Fanny Burney." *Literature Compass* 11, no. 3 (2014): 206–17.

Csengei, Ildiko. *Sympathy, Sensibility, and the Literature of Feeling in the Eighteenth Century*. London: Palgrave Macmillan, 2012.

Curran, Andrew S. *The Anatomy of Blackness: Science and Slavery in an Age of Enlightenment*. Baltimore, MD: Johns Hopkins University Press, 2011.

Curran, Stuart. "The I Altered." In *Romanticism and Feminism*, edited by Anne K. Mellor, 185–207. Indianapolis: Indiana University Press, 1988.

DeCoursey, Patricia J. "The Behavioral Ecology and Evolution of Biological Timing Systems." In *Chronobiology: Biological Timekeeping*, edited by Jay C. Dunlap, Jennifer J. Loros, and Patricia J. De Coursey, 27–65. Sunderland, MA: Sinauer, 2004.

DeLucia, JoEllen. *A Feminine Enlightenment: British Women Writers and the Philosophy of Progress, 1759–1820*. Edinburgh: Edinburgh University Press, 2015.

Delaporte, François. *Nature's Second Kingdom: Explorations in Vegetality in the Eighteenth Century*. Translated by by Arthur Goldhammer. Cambridge, MA: MIT Press, 1982.

Dolan, Elizabeth A. *Seeing Suffering in Women's Literature of the Romantic Era*. Burlington, VT: Ashgate, 2008.

Donovan, Julie. *Sydney Owenson, Lady Morgan and the Politics of Style*. Palo Alto, CA: Academica, 2009.

———. "Text and Textile in Sydney Owenson's *The Wild Irish Girl*." *Éire-Ireland* 43, nos. 3–4 (Fall–Winter 2008): 31–57.

Drayton, Richard. *Nature's Government: Science, Imperial Britain, and the "Improvement" of the World*. New Haven, CT: Yale University Press, 2000.

Drouin, Jean-Marc, and Bernadette Bensaude-Vincent. "Nature for the People." In *Cultures of Natural History*, edited by N. Jardine, J. A. Saecord, and E. C. Spary, 408–25. Cambridge: Cambridge University Press, 1996.

Drury, Joseph. "Haywood's Thinking Machines." *Eighteenth-Century Fiction* 21, no. 2 (Winter 2008–9): 201–28.

Egerton, Frank N. *Roots of Ecology: Antiquity to Haeckel*. Berkeley: University of California Press, 2012.

Ellis, Markman. *The Politics of Sensibility: Race, Gender, and Commerce in the Sentimental Novel*. Cambridge: Cambridge University Press, 1996.

Faflak, Joel, and Richard C. Sha. "Introduction: Feeling Romanticism." In *Romanticism and the Emotions*, 1–18. Cambridge: Cambridge University Press, 2014.

Fara, Patricia. *Erasmus Darwin: Sex, Science, and Serendipity*. Oxford: Oxford University Press, 2012.

Feingold, Richard. *Nature and Society: Later Eighteenth-Century Uses of the Pastoral and Georgic*. New Brunswick, NJ: Rutgers University Press, 1978.

Ferris, Ina. "Narrating Cultural Encounter: Lady Morgan and the Irish National Tale." *Nineteenth-Century Literature* 51, no. 3 (1996): 287–303.
Fletcher, Loraine. *Charlotte Smith: A Critical Biography*. New York: St. Martin's, 1998.
Foucault, Michel. *The Order of Things: An Archaeology of the Human Sciences*. New York: Vintage, [1966] 1994.
Fox, Christopher, Roy Porter, and Robert Wokler, eds. *Inventing Human Science: Eighteenth-Century Domains*. Oakland: University of California Press, 1995.
Freeman, Hannah Cowles. "Opium Use and Romantic Women's Poetry." *South Central Review* 29, nos. 1–2 (Spring–Summer 2012): 1–20.
Fulford, Tim. "Coleridge, Darwin, Linnaeus: The Sexual Politics of Botany." *Wordsworth Circle* 28, no. 3 (Summer 1997): 124–30.
———. "De Quincey's Literature of Power." *Wordsworth Circle* 31, no. 3 (Summer 2000): 158–64.
Fulford, Tim, Debbie Lee, and Peter J. Kitson. *Literature, Science, and Exploration in the Romantic Era: Bodies of Knowledge*. Cambridge: Cambridge University Press, 2004.
Gaukroger, Stephen. *The Collapse of Mechanism and the Rise of Sensibility: Science and the Shaping of Modernity, 1680–1760*. Oxford: Clarendon, 2010.
Gaull, Marilyn. "Clare and 'the Dark System.'" In *John Clare in Context*, edited by Hugh Haughton, Adam Phillips, and Geoffrey Summerfield, 279–94. Cambridge: Cambridge University Press, 1994.
Geoghegan, Patrick M. *The Irish Act of Union: A Study in High Politics, 1798–1801*. New York: St. Martin's, 1999.
George, Sam. *Botany, Sexuality, and Women's Writing, 1760–1830: From Modest Shoot to Forward Plant*. Manchester: Manchester University Press, 2007.
———. "'Not Strictly Proper for a Female Pen': Eighteenth-Century Poetry and the Sexuality of Botany." *Comparative Critical Studies* 2, no. 2 (2005): 191–210.
Gibson, Susannah. *Animal, Vegetable, Mineral? How Eighteenth-Century Science Disrupted the Natural Order*. Oxford: Oxford University Press, 2015.
Gigante, Denise. *Taste: A Literary History*. New Haven, CT: Yale University Press, 2005.
Gladstone, Hugh S. *Maria Riddell, The Friend of Burns*. Dumfries: Council of the Dumfriesshire & Galloway Natural History and Antiquarian Society, 1915.
Golinski, Jan. *British Weather and the Climate of Enlightenment*. Chicago: University of Chicago Press, 2007.
Goodman, Kevis. *Georgic Modernity and British Romanticism: Poetry and the Mediation of History*. Cambridge: Cambridge University Press, 2004.
Gould, Stephen Jay. *Time's Arrow, Time's Cycle: Myth and Metaphor in the Discovery of Geological Time*. Cambridge, MA; Harvard University Press, 1987.

Grainger, Margaret, ed. *The Natural History Prose Writings of John Clare*. Oxford: Oxford University Press, 1983.
Greene, Jack P. *Evaluating Empire and Confronting Colonialism in Eighteenth-Century Britain*. Cambridge: Cambridge University Press, 2013.
Grene, Marjorie and David Depew. *Philosophy of Biology: An Episodic History*. Cambridge: Cambridge University Press, 2004.
Griffiths, Devin S. "The Intuitions of Analogy in Erasmus Darwin's Poetics." *SEL* 51, no. 3 (Summer 2011): 645–65.
Groom, Nick. "William Henry Ireland: From Forgery to Fish'n'Chips." In *Cultures of Taste / Theories of Appetite: Eating Romanticism*, edited by Timothy Morton, 21–40. New York: Palgrave Macmillan, 2004.
Grove, Richard H. *Green Imperialism: Colonial Expansion, Tropical Island Edens, and the Origins of Environmentalism, 1600–1860*. Cambridge: Cambridge University Press, 1995.
Hall, S. Cailey. "The Wild Irish Girl Diet." *SEL: Studies in English Literature* 60, no. 3 (Summer 2020): 551–75.
Hartman, Geoffrey. *Wordsworth's Poetry, 1787–1814*. Cambridge, MA: Harvard University Press, [1964] 1987.
Hassler, Donald M. *The Comedian as the Letter D: Erasmus Darwin's Comic Materialism*. The Hague: Martinus Nijhoff, 1973.
Haughton, Hugh. "Progress and Rhyme: 'The Nightingale's Nest' and Romantic Poetry." In *John Clare in Context*, edited by Hugh Haughton, Adam Phillips, and Geoffrey Summerfield, 51–86. Cambridge: Cambridge University Press, 1994.
Haut, Asia. "Reading Flora: Erasmus Darwin's *The Botanic Garden*, Henry Fuseli's Illustrations, and Various Literary Responses." *Word & Image* 20, no. 4 (October–December 2004): 240–56.
Hilbish, Florence. *Charlotte Smith, Poet and Novelist, 1749–1806*. Philadelphia: n.p., 1911.
Holberton, Edward. "James Thomson's *The Seasons* and the Empire of the Seas." *Huntington Library Quarterly* 78, no. 1 (2015): 41–60.
Ingram, Allan, with Michelle Faubert. *Cultural Constructions of Madness in Eighteenth-Century Writing: Representing the Insane*. London: Palgrave Macmillan, 2005.
Jackson, Noel. "Rhyme and Reason: Erasmus Darwin's Romanticism." *Modern Language Quarterly* 70, no. 2 (June 2009): 171–94.
———. *Science and Sensation in Romantic Poetry*. Cambridge: Cambridge University Press, 2008.
Johnson, Claudia L. *Equivocal Beings: Politics, Gender, and Sentimentality in the 1790s: Wollstonecraft, Radcliffe, Burney, Austen*. Chicago: University of Chicago Press, 1995.

Jordanova, Ludmilla. *Sexual Visions: Images of Gender in Science and Medicine between the Eighteenth and Twentieth Centuries*. Madison: University of Wisconsin Press, 1989.

Juengel, Scott. "Countenancing History: Mary Wollstonecraft, Samuel Stanhope Smith, and Enlightenment Racial Science." *ELH* 68, no. 4 (2001): 897–927.

Kairoff, Claudia. *Anna Seward and the End of the Eighteenth Century*. Baltimore, MD: Johns Hopkins University Press, 2012.

———. "Samuel Johnson and Anna Seward: Solitude and Sensibility." In *Community and Solitude*, edited by Anthony Lee, 191–213. Lewisburg, PA: Bucknell University Press, 2019.

Kaul, Suvir. *Poems of Nation, Anthems of Empire: English Verse in the Long Eighteenth Century*. Charlottesville: University of Virginia Press, 2000.

Keith, Jennifer. *Poetry and the Feminine from Behn to Cowper*. Newark: University of Delaware Press, 2005.

———. "Poetry, Sentiment, and Sensibility." In *A Companion to Eighteenth-Century Poetry*, edited by Christine Gerrard, 127–41. New York: Blackwell, 2006.

Kelley, Theresa. *Clandestine Marriage: Botany and Romantic Culture*. Baltimore, MD: Johns Hopkins University Press, 2012.

———. "Romantic Exemplarity: Botany and 'Material' Culture." In *Romantic Science: The Literary Forms of Natural History*, edited by Noah Heringman, 223–54. Albany: State University of New York Press, 2003.

Kelly, Gary, ed. *Felicia Hemans: Selected Poetry, Prose, and Letters*. Peterborough, ONT: Broadview, 2002.

Kennedy, Deborah. "Wordsworth, Hemans, and the 'Literary Lady.'" *Victorian Poetry* 35, no. 3 (Fall 1997): 267–85.

Kerr, Heather. "Melancholy Botany: Charlotte Smith's Bioregional Poetic Imaginary." In *The Bioregional Imagination: Literature, Ecology, and Place*, 181–99. Athens: University of Georgia Press, 2012.

Kiberd, Declan. *Inventing Ireland: The Literature of the Modern Nation*. Cambridge, MA: Harvard University Press, [1995] 2002.

Kim, Elizabeth S. "Complicating 'Complicity/Resistance' in Janet Schaw's *Journal of a Lady of Quality*." *A/B: Auto/Biography Studies* 12, no. 2 (1997): 172–77.

King, Amy M. *Bloom: The Botanical Vernacular in the English Novel*. Oxford: Oxford University Press, 2007.

King-Hele, Desmond. *Erasmus Darwin: A Life of Unequalled Achievement*. London: Giles del la Mare, 1999.

———. *Erasmus Darwin and the Romantic Poets*. New York: St. Martin's, 1986.

———. "Erasmus Darwin, Man of Ideas and Inventor of Words." *Notes and Records of the Royal Society of London* 42, no. 2 (July 1988): 140–80.

Kirkpatrick, Kathryn. "Introduction." In *The Wild Irish Girl: A National Tale* by Sydney Owenson, Lady Morgan, edited with introduction and notes by Kathryn Kirkpatrick, vii–xvii. Oxford: Oxford University Press, 1999.

Kitson, Peter J. *Romantic Literature, Race, and Colonial Encounter*. London: Palgrave Macmillan, 2007.

———. "The Strange Case of Dr. White and Mr. De Quincey: Manchester, Medicine and Romantic Theories of Biological Racism." *Romanticism* 17, no. 3 (2011): 278–87.

Knight, David. *Ordering the World: A History of Classifying Man*. London: Burnett, 1981.

Koerner, Lisbet. *Linnaeus: Nature and Nation*. Cambridge, MA: Harvard University Press, 1999.

———. "Purposes of Linnaean Travel: A Preliminary Research Report." In *Visions of Empire: Voyages, Botany, and Representations of Nature*, edited by David Philip Miller and Peter Hanns Reill, 117–52. Cambridge: Cambridge University Press, 1996.

Krueger, Misty, ed. *Transatlantic Women Travelers, 1688–1843*. Lewisburg, PA: Bucknell University Press, 2021.

Labbe, Jacqueline. *Charlotte Smith: Romanticism, Poetry, and the Culture of Gender*. Manchester: Manchester University Press, 2003.

———. "Pathological Sensibility." *Women's Writing* 23, no. 3 (2016): 354–65.

———. *Romantic Visualities: Landscape, Gender, and Romanticism*. Houndmills: Macmillan, 1998.

———. "'Transplanted into More Congenial Soil': Footnoting the Self in the Poetry of Charlotte Smith." In *Ma(r)king the Text: The Presentation of Meaning on the Literary Page*, edited by Joe Bray, Miriam Handley, and Anne C. Henry, 71–86. Aldershot: Ashgate, 2000.

———. *Writing Romanticism: Charlotte Smith and William Wordsworth, 1784–1807*. New York: Palgrave Macmillan, 2011.

Lamb, Jonathan. *The Evolution of Sympathy in the Long Eighteenth Century*. London: Pickering & Chatto, 2009.

Landry, Donna. "Green Languages? Women Poets as Naturalists in 1653 and 1807." In *Forging Connections: Women's Poetry from the Renaissance to Romanticism*, edited by Anne K. Mellor, Felicity Nussbaum, and Jonathan F. S. Post, 39–61. San Marino, CA: Huntington Library, 2002.

Leader, Zachary. *Revision and Romantic Authorship*. Oxford: Clarendon, 1996.

Lee, Wendy Anne. *Failures of Feeling: Insensibility and the Novel*. Stanford, CA: Stanford University Press, 2019.

Leighton, Angela. *Victorian Women Poets: Writing against the Heart*. Charlottesville: University of Virginia Press, 1992.

Levinson, Marjorie. "A Motion and a Spirit: Romancing Spinoza." *Studies in Romanticism* 46, no. 4 (2007): 367–408.

———. "Of Being Numerous." *Studies in Romanticism* 49, no. 4 (2010): 633–57.
Lew, Joseph. "Sidney Owenson and the Fate of Empire." *Keats-Shelley Journal* 39 (1990): 39–65.
List, Julia. "Sometimes a Stamen Is Only a Stamen: Sexuality, Women, and Darwin's *Loves of the Plants*." *Nineteenth-Century Contexts* 32, no. 3 (2010): 199–218.
Livesay, Daniel. "The Decline of Jamaica's Interracial Households and the Fall of the Planter Class, 1733–1823." In *Rethinking the Fall of the Planter Class*, edited by Christer Petley, 107–23. New York: Routledge, 2018.
Looser, Devoney. "Why I'm Still Writing Women's Literary History." *Minnesota Review* 71–72 (Winter–Spring 2009): 220–27.
Macfarlane, Robert. *Original Copy: Plagiarism and Originality in Nineteenth-Century Literature*. Oxford: Oxford University Press, 2007.
Macnaghten, Angus. *Burns's Mrs. Riddell: A Biography*. Peterhead: Volutrna, 1975.
Mahood, M. M. *The Poet as Botanist*. Cambridge: Cambridge University Press, 2008.
Makdisi, Saree. *Romantic Imperialism: Universal Empire and the Culture of Modernity*. Cambridge: Cambridge University Press, 1998.
Mazzeo, Tilar J. *Plagiarism and Literary Property in the Romantic Period*. Philadelphia: University of Pennsylvania Press, 2007.
McCarthy, William. *Anna Letitia Barbauld: Voice of the Enlightenment*. Baltimore, MD: Johns Hopkins University Press, 2008.
McGann, Jerome. *The Poetics of Sensibility: A Revolution in Literary Style*. Oxford: Oxford University Press, 1996.
———. *The Romantic Ideology: A Critical Investigation*. Chicago: University of Chicago Press, 1983.
McGrath, Brian. "Thomas De Quincey and the Language of Literature: Or, On the Necessity of Ignorance." *SEL: Studies in English Literature* 47, no. 4 (Autumn 2007): 847–62.
McKillop, Alan Dugald. *The Background of Thomson's Seasons*. Minneapolis: University of Minnesota Press, 1942.
McKusick, James. "Beyond the Visionary Company: John Clare's Resistance to Romanticism." In *John Clare in Context*, edited by Hugh Haughton, Adam Phillips, and Geoffrey Summerfield, 221–37. Cambridge: Cambridge University Press, 1994.
McLaverty, James. *Pope, Print, and Meaning*. Oxford: Oxford University Press, 2001.
McNeil, Maureen. *Under the Banner of Science: Erasmus Darwin and His Age*. Manchester: Manchester University Press, 1987.
Mellor, Anne K. "The Baffling Swallow: Gilbert White, Charlotte Smith, and the Limits of Natural History." *Nineteenth-Century Contexts* 31, no. 4 (December 2009): 299–309.

———. *Romanticism and Gender*. New York: Routledge, 1993.

———. "Were Women Writers 'Romantics'?" *MLQ: Modern Language Quarterly* 62, no. 4 (2001): 393–405.

Melnyk, Julie. "William Wordsworth and Felicia Hemans." In *Fellow Romantics: Male and Female British Writers, 1790–1835*, edited by Beth Lau, 139–58. Burlington, VT: Ashgate, 2009.

Menely, Tobias. "Ecologies of Time." In *Time and Literature*, edited by Thomas M. Allen, 85–100. Cambridge: Cambridge University Press, 2018.

Merrett, Robert James. "The Culinary Art of Eighteenth-Century Women Cookbook Authors." In *Women, Popular Culture, and the Eighteenth Century*, edited by Tiffany Potter, 115–32. Toronto: University of Toronto Press, 2012.

Mills, Catherine. *Biopolitics*. New York: Routledge, 2018.

Mitchell, Robert. "Cryptogamia." *European Romantic Review* 21, no. 5 (October 2010): 631–51.

———. *Experimental Life: Vitalism in Romantic Science and Literature*. Baltimore, MD: Johns Hopkins University Press, 2013.

Moore, Lisa L. "Acts of Union: Sexuality and Nationalism, Romance and Realism in the Irish National Tale." *Cultural Critique* 44 (Winter 2000): 114–44.

Mossner, Ernest Campbell. *The Life of David Hume*. 2nd ed. Oxford: Clarendon, [1954] 2001.

Mullan, John. "Hypochondria and Hysteria: Sensibility and the Physicians." *Eighteenth Century: Theory and Interpretation* 25, no. 2 (1984): 141–74.

———. *Sentiment and Sociability: The Language of Feeling in the Eighteenth Century*. Oxford: Clarendon, 1988.

Murphy, Patricia. *In Science's Shadow: Literary Constructions of Late Victorian Women*. Columbia: University of Missouri Press, 2006.

Myrone, Martin. "Gothic Romance and the Quixotic Hero: A Pageant for Henry Fuseli in 1783." Tate Papers, Spring 2004. https://www.tate.org.uk/research/publications/tate-papers/01/gothic-romance-and-the-quixotic-hero-a-pageant-for-henry-fuseli-1783 (accessed 23 June 2021).

———. *Henry Fuseli*. Princeton, NJ: Princeton University Press, 2001.

Nersessian, Anahid. "Empire and Attachment: A Transnational Tale." *European Romantic Review* 22, no. 3 (June 2011): 339–46.

Newcomer, James. *Lady Morgan: The Novelist*. Lewisburg, PA: Bucknell University Press, 1990.

Nicholson, Michael. "*Fugitive Pieces*: Walpole, Byron, and Queer Time." *Eighteenth Century: Theory and Interpretation* 60, no. 2 (2019): 139–62.

Nie, Michael de. *The Eternal Paddy: Irish Identity and the British Press, 1798–1882*. Madison: University of Wisconsin Press, 2004.

North, Julian. "De Quincey and the Inferiority of Women." *Romanticism* 17, no. 3 (2011): 327–39.

Novak, Maximillian E., and Anne Mellor. "Introduction." In *Passionate Encounters in a Time of Sensibility*, edited by Novak and Mellor, 11–26. Newark: University of Delaware Press, 2000.
Nussbaum, Felicity A. *The Limits of the Human: Fictions of Anomaly, Race, and Gender in the Long Eighteenth Century*. Cambridge: Cambridge University Press, 2003.
———. *Torrid Zones: Maternity, Sexuality, and Empire in Eighteenth-Century English Narratives*. Baltimore, MD: Johns Hopkins University Press, 1995.
Oldroyd, David. *Thinking about the Earth: A History of Ideas in Geology*. Cambridge, MA: Harvard University Press, 1996.
Ottum, Lisa. "'Shallow' Estates and the 'Deep' Wild: The Landscapes of Charlotte Smith's Fiction." *Tulsa Studies in Women's Literature* 34, no. 2 (Fall 2015): 249–72.
Ottum, Lisa, and Seth T. Reno. "Introduction." In *Wordsworth and the Green Romantics: Affect and Ecology in the Nineteenth Century*, edited by Ottum and Reno, 1–27. Lebanon: University of New Hampshire Press, 2016.
Oxley, Valerie. *Botanical Illustration*. Ramsbury: Crowood, 2008.
Packham, Catherine. "The Science and Poetry of Animation: Personification, Analogy, and Erasmus Darwin's *Loves of the Plants*." *Romanticism: The Journal of Romantic Culture and Criticism* 10, no. 2 (2004): 191–208.
Page, Judith W., and Elise L. Smith. *Women, Literature, and the Domesticated Landscape: England's Disciples of Flora, 1780–1870*. Cambridge: Cambridge University Press, 2011.
Palmeri, Frank. *State of Nature, Stages of Society: Enlightenment Conjectural History and Modern Social Discourse*. New York: Columbia University Press, 2016.
Parrish, Susan Scott. *American Curiosity: Cultures of Natural History in the Colonial British Atlantic World*. Chapel Hill: University of North Carolina Press, 2006.
———. "Science, Nature, Race." In *The Atlantic World: 1450–1850*, edited by Nicholas Canny and Philip Morgan, 463–79. Oxford: Oxford University Press, 2011.
Pascoe, Judith. "Female Botanists and the Poetry of Charlotte Smith." In *Re-Visioning Romanticism: British Women Writers, 1776–1837*, edited by Carol Shiner Wilson and Joel Haefner, 193–209. Philadelphia: University of Pennsylvania Press, 1994.
Perovic, Sanja. *The Calendar in Revolutionary France: Perceptions of Time in Literature, Culture, and Politics*. Cambridge: Cambridge University Press, 2012.
Peterson, Linda H. *Becoming a Woman of Letters: Myths of Authorship and Facts of the Victorian Market*. Princeton, NJ: Princeton University Press, 2009.
Petley, Christer. "Gluttony, Excess, and the Fall of the Planter Class in the British Caribbean." In *Rethinking the Fall of the Planter Class*, edited by Christer Petley, 85–106. London: Routledge, 2017.

Pinch, Adela. *Strange Fits of Passion: Epistemologies of Emotion, Hume to Austen.* Stanford, CA: Stanford University Press, 1996.

Pittock, Murray G. H. *Poetry and Jacobite Politics in Eighteenth-Century Britain and Ireland.* Cambridge: Cambridge University Press, 1994.

Plasa, Carl. *Slaves to Sweetness: British and Caribbean Literatures of Sugar.* Liverpool: Liverpool University Press, 2009.

Pliny, the Elder. *Natural History.* 10 vols. Translated by H. Rackham. Cambridge, MA: Harvard University Press, 1938–62.

Porter, Dahlia. "Formal Relocations: The Method of Southey's Thalaba the Destroyer (1801)." *European Romantic Review* 20.5 (Dec. 2009): 671–79.

———. "From Nosegay to Specimen Cabinet: Charlotte Smith and the Labour of Collecting." In *Charlotte Smith in British Romanticism*, edited by Jacqueline Labbe, 29–44. London: Pickering & Chatto, 2008.

———. *Science, Form, and the Problem of Induction in British Romanticism.* Cambridge University Press, 2018.

———. "Scientific Analogy and Literary Taxonomy in Darwin's *Loves of the Plants.*" *European Romantic Review* 18, no. 2 (April 2007): 213–21.

Porter, Roy. "'All Madness for Writing': John Clare and the Asylum." In *John Clare in Context*, edited by Hugh Haughton, Adam Phillips, and Geoffrey Summerfield, 259–78. Cambridge: Cambridge University Press, 1994.

———. *Madness: A Brief History.* Oxford: Oxford University Press, 2002.

Powell, Martyn J. *The Politics of Consumption in Eighteenth-Century Ireland.* Basingstoke: Palgrave Macmillan, 2005.

Powell, Rosalind. "Linnaeus, Analogy, and Taxonomy: Botanical Naming and Categorization in Erasmus Darwin and Charlotte Smith." *Philological Quarterly* 95, no. 1 (2016): 101–24.

Pratt, Mary Louise. *Imperial Eyes: Travel Writing and Transculturation.* London: Routledge, 1992.

Priestman, Martin. *The Poetry of Erasmus Darwin: Enlightened Spaces, Romantic Times.* Farnham: Ashgate, 2013.

———. "'The Progress of Society?' Darwin's Early Drafts for *The Temple of Nature.*" In *The Genius of Erasmus Darwin*, edited by C. U. M. Smith and Robert Arnott, 307–19. Farnham: Ashgate, 2005.

———. *Romantic Atheism: Poetry and Freethought, 1780–1830.* Cambridge: Cambridge University Press, 1999.

Reill, Peter Hans. *Vitalizing Nature in the Enlightenment.* Berkeley: University of California Press, 2005.

Reiman, Donald H. "Introduction." In *The Botanic Garden: A Poem in Two Parts*, n.p. New York: Garland, 1978.

Richardson, Alan. "Romanticism and the Colonization of the Feminine." In *Romanticism and Feminism*, edited by Anne K. Mellor, 13–25. Bloomington: Indiana University Press, 1988.

Riskin, Jessica. *Science in the Age of Sensibility: The Sentimental Empiricists of the French Enlightenment.* Chicago: University of Chicago Press, 2002.

Robinson, Daniel. "*The River Duddon* and Wordsworth, Sonneteer." In *The Oxford Handbook of William Wordsworth*, edited by Richard Gravil and Daniel Robinson, n.p. Oxford: Oxford University Press, 2015.

Rodgers, Nini. *Ireland, Slavery, and Anti-Slavery: 1612–1865.* New York: Palgrave Macmillan, 2007.

Roger, Jacques. *Buffon: A Life in Natural History.* Translated by Sarah Lucille Bonnefoi. Ithaca, NY: Cornell University Press, 1997.

Ross, Catherine. "'Twin Labourers and Heirs of the Same Hopes': The Professional Rivalry of Humphry Davy and William Wordsworth." In *Romantic Science*, edited by Noah Heringman, 23–52. Albany: State University of New York Press, 2003.

Ross, Marlon B. *The Contours of Masculine Desire: Romanticism and the Rise of Women's Poetry.* Oxford: Oxford University Press, 1989.

Rothstein, Eric. *Restoration and Eighteenth-Century Poetry, 1660–1780.* Vol. 3 of *The Routledge History of English Poetry.* R. A. Foakes, general editor. London: Routledge & Kegan Paul, 1981.

Rousseau, G. S. "Nerves, Spirits and Fibres: Towards Defining the Origins of Sensibility." In *Studies in the Eighteenth Century 3*, edited by R. F. Brissenden and J. E. Eade, 137–57. Toronto: University of Toronto Press, 1976.

Ruston, Sharon. *Creating Romanticism: Case Studies in the Literature, Science, and Medicine of the 1790s.* London: Palgrave Macmillan, 2013.

Ruwe, Donelle R. "Charlotte Smith's Sublime: Feminine Poetics, Botany and 'Beachy Head.'" *Prism(s): Essays in Romanticism* 7 (1999): 117–32.

Sachs, Jonathan. *The Poetics of Decline in British Romanticism.* Cambridge: Cambridge University Press, 2018.

Sagal, Anna K. *Botanical Entanglements: Women, Natural Science, and the Arts in Eighteenth-Century England.* Charlottesville: University of Virginia Press, 2022.

Salaman, Redcliffe. *The Social History and Influence of the Potato.* Revised by J. G. Hawkins. Cambridge: Cambridge University Press, 1985.

Sambrook, James. *James Thomson, 1700–1748: A Life.* Oxford: Clarendon, 1991.

Schiebinger, Londa. "Gender and Natural History." In *Cultures of Natural History*, edited by N. Jardine, J. A. Secord, and E. C. Spary, 163–77. Cambridge: Cambridge University Press, 1996.

———. *Nature's Body: Gender in the Making of Modern Science.* New York: Routledge, [1993] 2004.

———. *Plants and Empire: Colonial Bioprospecting in the Atlantic World.* Cambridge, MA: Harvard University Press, 2004.

———. "The Private Life of Plants: Sexual Politics in Carl Linnaeus and Erasmus Darwin." In *Science and Sensibility: Gender and Scientific Enquiry, 1780–1945*, edited by Marina Benjamin, 121–43. New York: Basil Blackwell, 1991.

———. "Scientific Exchange in the Eighteenth-Century Atlantic World." In *Soundings in Atlantic History: Latent Structures and Intellectual Currents, 1500–1830*, edited by Bernard Bailyn and Patricia L. Denault, 294–328. Cambridge, MA: Harvard University Press, 2009.

Schmidgen, Wolfram. *Exquisite Mixture: The Virtues of Impurity in Early Modern England*. Philadelphia: University of Pennsylvania Press, 2012.

Schuller, Kyla. *The Biopolitics of Feeling: Race, Sex, and Science in the Nineteenth Century*. Durham, NC: Duke University Press, 2018.

Seaton, Beverley. *The Language of Flowers: A History*. Charlottesville: University of Virginia Prss, 1995.

Sherry, Vincent. *Modernism and the Reinvention of Decadence*. Cambridge: Cambridge University Press, 2015.

Showalter, Elaine. *The Female Malady: Women, Madness, and English Culture, 1830–1980*. New York: Pantheon, 1985.

Shteir, Ann B. *Cultivating Women, Cultivating Science: Flora's Daughters and Botany in England, 1760 to 1860*. Baltimore, MD: Johns Hopkins University Press, 1996.

Shteir, Ann. "'She comes!—the Goddess!' Narrating Nature in Erasmus Darwin's *The Botanic Garden*." In *Fact and Fiction: Literary and Scientific Cultures in Germany and Britain*, edited by Christine Lehleiter, 73–96. Toronto: University of Toronto Press, 2016.

Siskin, Clifford. *The Historicity of Romantic Discourse*. Oxford: Oxford University Press, 1988.

Sitter, John. *Literary Loneliness in Mid-Eighteenth-Century England*. Ithaca, NY: Cornell University Press, 1982.

Sloan, Phillip. "The Gaze of Natural History." In *Inventing Human Science: Eighteenth-Century Domains*, edited by Christopher Fox, Roy Porter, and Robert Wokler, 112–51. Berkeley: University of California Press, 1995.

Somers, David E. "The Physiology and Molecular Bases of the Plant Circadian Clock." *Plant Physiology* 121 (September 1999): 9–20.

Spacks, Patricia Meyer. *The Varied God: A Critical Study of Thomson's* The Seasons. Berkeley: University California of Press, 1959.

Stafford, Barbara. "Images of Ambiguity: Eighteenth-Century Microscopy and the Neither/Nor." In *Visions of Empire: Voyages, Botany, and Representations of Nature*, edited by David Philip Miller and Peter Hanns Reill, 230–57. Cambridge: Cambridge University Press, 1996.

Stillinger, Jack. *Multiple Authorship and the Myth of Solitary Genius*. Oxford: Oxford University Press, 1991.

Storey, Mark. "Clare and the Critics." In *John Clare in Context*, edited by Hugh Haughton, Adam Phillips, and Geoffrey Summerfield, 28–50. Cambridge: Cambridge University Press, 1994.

Sussman, Charlotte. *Consuming Anxieties: Consumer Protest, Gender, and British Slavery, 1713–1833*. Stanford, CA: Stanford University Press, 2000.

Sweet, John Wood. *Bodies Politic: Negotiating Race in the American North, 1730–1830.* Philadelphia: University of Pennsylvania Press, 2006.

Taylor, Jesse Oak. *The Sky of Our Manufacture: The London Fog in British Fiction from Dickens to Woolf.* Charlottesville: University of Virginia Press, 2016.

Taylor, Joseph. *The Complete Weather Guide.* London: Printed for John Harding, 1812.

Tessone, Natasha. "Displaying Ireland: Sydney Owenson and the Politics of Spectacular Antiquarianism." *Éire-Ireland* 37, nos. 3–4 (2002): 169–86.

Teute, Fredrika J. "The Loves of the Plants; or, The Cross-Fertilization of Science and Desire at the End of the Eighteenth Century." *Huntington Library Quarterly* 63, no. 3 (2000): 319–45.

Thompson, Carl. "Women Travelers, Romantic-Era Science and the Banksian Empire." *Notes & Records* 73 (2019): 431–55.

Thomson, Keith. *Before Darwin: Reconciling God and Nature.* New Haven, CT: Yale University Press, 2005.

Tibble, J. W. and A., eds. *The Prose of John Clare.* London: Routledge, 1951.

Tobin, Beth Fowkes. *Colonizing Nature: The Tropics in British Arts and Letters, 1760–1820.* Philadelphia: University of Pennsylvania Press, 2005.

Tracy, Thomas. "The Mild Irish Girl: Domesticating the National Tale." *Éire-Ireland* 39, nos. 1–2 (Spring–Summer 2004): 81–109.

Trott, Nicola. "Wordsworth's Loves of the Plants." In *1800: The New Lyrical Ballads*, edited by Nicola Trott and Seamus Perry, 141–68. London: Palgrave Macmillan, 2001.

Trumpener, Katie. "National Character, Nationalist Plots: National Tale and Historical Novel in the Age of Waverley, 1806–1830." *ELH* 60 (Fall 1993): 685–731.

Turner, Katherine. *British Travel Writers in Europe, 1750–1800: Authorship, Gender, and National Identity.* Aldershot: Ashgate, 2001.

Uglow, Jenny. *The Lunar Men: The Friends Who Made the Future, 1730–1810.* New York: Farrar, Straus & Giroux, 2002.

Van Sant, Ann Jessie. *Eighteenth-Century Sensibility and the Novel: The Senses in Social Context.* Cambridge: Cambridge University Press, 1993.

Weaver, Karol K. *Medical Revolutionaries: The Enslaved Healers of Eighteenth-Century Saint Domingue.* Urbana: University of Illinois Press, 2006.

Whale, John. "De Quincey and Men (of Letters)." In *Thomas De Quincey: New Theoretical and Critical Directions*, edited by Robert Morrison and Daniel Sanjiv Roberts, 81–98. New York: Routledge, 2008.

Wheeler, Michael. *The Old Enemies: Catholic and Protestant in Nineteenth-Century English Culture.* Cambridge: Cambridge University Press, 2006.

Wheeler, Roxann. *The Complexion of Race: Categories of Difference in Eighteenth-Century British Culture.* Philadelphia: University of Pennsylvania Press, 2000.

Whippo, Craig W. and Roger P. Hangarter. "The 'Sensational' Power of Movement in Plants: A Darwinian System for Studying the Evolution of Behavior." *American Journal of Botany* 96, no. 12 (2009): 2115–27.

Whitehead, James. *Madness and the Romantic Poet: A Critical History*. Oxford: Oxford University Press, 2017.

Willan, Claude. "The Proper Study of Mankind in Pope and Thomson." *ELH* 84 (2017): 63–90.

Williams, Jane. *The Literary Women of England*. London: Saunders, Otley, 1861.

Wilson, Erin. "The End of Sensibility: The Nervous Body in the Early Nineteenth Century." *Literature & Medicine* 30, no. 2 (Fall 2012): 276–91.

Wilson, Kathleen. *The Island Race: Englishness, Empire, and Gender in the Eighteenth Century*. New York: Routledge, 2003.

Wood, Gillen D'Arcy. *Tambora: The Eruption That Changed the World*. Princeton, NJ: Princeton University Press, 2014.

Wood, Gillen. "The Female Penseroso: Anna Seward, Sociable Poetry, and the Handelian Consensus." *MLQ: Modern Language Quarterly* 67, no. 4 (December 2006): 451–77.

Yahav, Amit S. *Feeling Time: Duration, the Novel, and Eighteenth-Century Sensibility*. Philadelphia: University of Pennsylvania Press, 2018.

Yolton, John W. *Thinking Matter: Materialism in Eighteenth-Century Britain*. New York: Blackwell, 1983.

Young, Robert J. C. *Colonial Desire: Hybridity in Theory, Culture, and Race*. London: Routledge, 1995.

INDEX

Page numbers in italics refer to illustrations.

abolitionist movement, 88, 101, 104–5, 116
Act of Union (1800), 30–31, 110, 112, 117, 118
Addison, Joseph, 39, 41
aesthetics of Romantic-era literature: about, 1–15, 19, 22, 32, 188, 206–7; De Quincey on, 190, 192, 197, 204; Jackson on, 174; in national and natural hybridity, 111; Porter on, 232n13; Riddell and, 106; Seward on, 173; Smith on, 142, 143, 144; standardization of, 28; of West Indian botany, 95; of Wordsworth, 188, 196. *See also* Romanticism
Ahern, Stephen, 12
Aikin, John: *Calendar of Nature*, 27, 50; *Essay on the Application of Natural History to Poetry*, 7, 9–10, 27, 38, 137, 142; on Linnaean botany, 27; on originality, 7–8; on *The Seasons*, 37–38; on truth, 8, 142, 213n91
Aikin, Lucy, 23, 27
Allen, Matthew, 180
American Revolution, 88, 101
Amulet, The (Hemans), 79
anthropomorphism, 67, 68, 70–71, 82, 115, 120, 123
anthropophages, 113, 115
Anti-Jacobin Review and Magazine (publication), 19, 159
Archaeologia (publication), 113
Arctic Zoology (Pennant), 38
Arum maculatum, 172
Athenaeum (publication), 116
Autumn. See Seasons, The (Thomson)

backwardness of spring, 44, 47–53, 216n41, 218n66. *See also* progress vs. decline, in British discourse

"Backwardness of Spring Accounted For, 1772, The" (anonymous), 35, 53–54
"Backwardness of the Present Spring Accounted For, May 5th, 1782, The" (Mundy), 54, 217n62
"Backward Spring, The" (Clare), 58
backward time, 44–47, 216n30. *See also* progress vs. decline, in British discourse
Baillie, Joanna, 192
Banks, Joseph, 44, 158
Barbauld, Anna: on Darwin's *Botanic Garden*, 65; Hazlitt on, 192; *Lessons for Children*, 50; "On the Backwardness of Spring," 47, 48–50, 217n44; on women writers, 238n23; writings on sensibility, 5, 21
Barker-Benfield, G. J., 11, 12
Barnard, Teresa, 166
Barrell, John, 45, 204
Bartram, William, 17–18
Bate, Jonathan, 180
"Beachy Head" (Smith), 142, 152, 222n54
Beachy Head, Fables, and Other Poems (Smith), 72, 77, 135
Beddoes, Thomas, 16
Behrendt, Stephen, 200–201
Bergstrom, Carson, 38
Bewell, Alan, 25
Bhabha, Homi, 230n44, 231n58
Biographia Literaria (Coleridge), 17, 18, 174
biological racism. *See* scientific racism
biological sexism. *See* scientific sexism
biological transformations, 87
Blake, William, 220n25, 234n2
Blick, Fred, 174
Bobey, John, 102–3

Botanical Arrangement of All the Vegetables Naturally Growing in Great Britain (Withering), 26, 75–76, 140
botanical classification system, 4, 25–26, 32, 55, 138, 187, 225n29. *See also* Linnaeus, Carl; sexual system of plants
botanical clock. *See* floral clock theory
Botanical Dictionary (Milne), 75
botanical writings, 9–10, 25–28. *See also names of specific authors*; poet-naturalists; scientific literature
Botanic Garden, The (Darwin), 4, 9, 10, 14, 16
botanomania, 155
Braun, Heather, 127
breadfruit, 227n54
Britain. *See* United Kingdom
British Association for the Advancement of Science (BAAS), 190
British Critic (publication), 1
Brown, Creighton Nicholas, 111
Browne, Patrick, 99
Browning, Elizabeth Barrett, 208
Buffon, Georges-Louis Leclerc, comte de, 36, 52, 57, 65, 87, 89, 91, 94–96, 102–4
Burke, Edmund, 114
Burnet, Thomas, 59, 218n83
Burns, Andrew, 116
Burns, Robert, 89–90, 105
Burwick, Frederick, 176

Calendar of Flora (Stillingfleet), 50
Calendar of Nature (Aikin), 27, 50
Calvinism, 172–73
Campbell, Mary, 112
Candolle, Augustin-Pyramus de, 27, 187, 219n6
cannibalism, colonial, 111, 112–15
Canning, George, 159–60
Caribbean colonies of Britain, 87–88, 94–105
Cary, Henry F., 179
cashew tree, 99–100, 226n39
Cashew Tree and Butterfly Metamorphosis (Merian), 97

Casid, Jill H., 91
cassava, 101–2
Catholicism, 110, 112, 117, 126, 221n33
Chandler, James, 12
Chiles, Katy L., 96
chronobiology, 61–62, 67, 69
circadian rhythms, 61–83, 220n24
circannual rhythms, 35–60, 62, 220n24
Clare, John, 32, 47, 58–60, 78, 99, 177–81, 237n68
class consciousness, 128–30, 230n39
climate and sensibility, 35–44, 215n17, 215n19. *See also* frigid zone; torrid zone
Coleman, Deirdre, 96
Coleridge, Samuel Taylor: *Biographia Literaria*, 17, 18, 174; on Darwin's originality, 15–16; on Darwin vs. Wordsworth, 1; drug addiction of, 168; *Lyrical Ballads*, 1, 5, 175
Colley, Linda, 126
colonial cannibalism, 111, 112–15
colonialism, 30–31, 87–88, 94–105, 125–26, 128–30, 226n50
conjectural history, 36, 44–47
Conjectures on Original Composition (Young), 156
Conserva aegagropila, 137
Conversations Introducing Poetry (Smith), 31, 133, 135
Cook, James, 157, 170
copyright laws, 237n1
Cowper, William, 5, 16, 31–32, 135, 169–72
Cox, James, 61
Crabbe, George, 5
"Crazy Kate," 170, 171
Cruikshank, Isaac, 117, 118
Curran, Stuart, 12, 13

Darwin, Charles, 4, 14–15, 165, 205
Darwin, Erasmus: biography of, 14; on botanical clock, 64, 66–71; botanical writings by, 9–10, 29; *The Botanic Garden*, 4, 9, 10, 14, 16, 64; "The Dream," 106, 107, 108; *The Economy of Vegetation*, 4, 10, 57, 159, 170; on irritability

of plants, 221n39; literary criticisms of, 9, 14, 65; *The Loves of the Plants*, 4, 10, 15, 16, 64, 115; on madness, 193–94; on nerve theory and lack of sensation, 19, 21, 24, 31, 156, 160–67; on novelty of poetry, 9; "Ode to the River Darwent," 106, 107–8; originality of, 15–16, 212n63; *Phytologia*, 70, 74, 81, 115, 166; "A Poetical Courtship," 107; Riddell and, 105–6; Smith and, 137–39; as target of literary critics, 1, 15–24; *The Temple of Nature*, 17, 56, 69; *Zoonomia*, 16, 31, 98, 156, 162–65, 194, 224n15. See also scientific literature
Darwin, Francis, 107
Davy, Humphry, 70, 187, 237n1
degeneration, 87–88
deistic materialism, 221nn32–34
Delany, Mary, 182–84
"Delights of Nature, The" (Linnaeus), 55, 218n67
De Mairan, Jean Jacques d'Ortous, 61–62
De Quincey, Thomas: on aesthetics of Romantic-era literature, 2, 190, 192, 197, 204; "False Distinctions," 197, 202, 206; "Joan of Arc," 204; "Letters to a Young Man Whose Education Has Been Neglected," 192, 196–97, 206; literary criticism by, 184–85, 192–207; on "literature of knowledge" vs. "literature of power," 4, 9, 17, 19, 185, 189, 190, 192–96, 202–7; on madness, 193–94; on women in literature, 197–99; on women in science, 194, 195–96, 197, 205; "The Works of Alexander Pope," 202
Descartes, René, 68
"Deserted Village, The" (Goldsmith), 52
determinism, 88
"Dial of Flowers, The" (Hemans), 79, 223n72
Discourse on Method (Descartes), 68
disease, 101
Dissertation upon Genius (Sharpe), 156
"Dream, The" (Darwin), 106, 107, 108
drug addiction, 167–68

dualism, 68
Dyce, Alexander, 200

Economy of Vegetation, The (Darwin), 4, 10, 57, 159, 170
Edgeworth, Richard, 16
Edinburgh Magazine (publication), 1
egotism, 154, 156, 172–76, 178, 191, 236n54
Elegiac Sonnets (Smith), 74, 136, 138
Elegy on Captain Cook (Seward), 157
"Elegy on the Death of Captain J. Woodley" (Riddell), 108
Elements of Botany (Rose), 73
Elements of Criticism (Kames), 19
Ellis, John, 149
Ellis, Markman, 12
"Essay, Supplementary to the Preface" (Wordsworth), 202–3
Essay on Man (Pope), 142
Essay on the Application of Natural History to Poetry (Aikin), 7, 9–10, 27, 38, 137, 142
Essay on the Classification of the Insane (Allen), 180
Essay on the History of Civil Society (Ferguson), 45
Essay on the Modifications of Clouds (Howard), 174
Essay on the Probability of Sensation in Vegetables, An (Tupper), 78
Essay Towards a Natural History of the Corallines (Ellis), 149
"Evening Walk by the Sea Side, An" (Smith), 151
evil eye, 230n44

Fable of the Bees (Mandeville), 35, 46, 218n69
"False Distinctions" (De Quincey), 197, 202, 206
Family Herbal (Hill), 178
famine, 113, 116–20
Fellowes, Robert, 160
femininity, 11–13, 204. See also gender
Ferguson, Adam, 45
Ferris, Ina, 127

Fisher, Elizabeth, 199
Flora (goddess), 55–56, 138, 139, 153, 232n14
"Flora" (poem by Smith), 31, 72, 133–43, 146, 147, 152–53
floral clock theory, 222n58; Darwin on, 66–71; Delaporte on, 222n49; Grainger on, 62–64; Hemans on, 79–83; by Linnaeus, 29–30, 60, 63, 221n28; Smith on, 71–72, 74–75. *See also* mechanical descriptions of plants
Flora Scotica (Lightfoot), 75
Foucault, Michel, 234n2
"Fourth Eclogue" (Virgil), 51–52
French peasantry, 228n18
French Revolution, 9, 23, 55, 73, 88, 114
frigid zone, 29, 36, 37, 41, 42–44, 47, 49, 52, 59, 216n33. *See also* climate and sensibility
Fucus natans, 137
Fulford, Tim, 204
Fuseli, Henry, 170–72

Gaull, Marilyn, 182
gender: botany and, 133, 141–42, 178, 213; literature and, 3, 197–207, 209n14; Riddell on, 90; science and, 8–9, 23, 25, 26, 195–97. *See also* femininity; masculinity; sexual system of plants
Genera Plantarum (Linnaeus), 26
Gentleman's Magazine, 25, 48, 159
Geoghegan, Patrick, 110
George, Sam, 25, 127, 140, 229n32
georgic poetry: by Darwin, 210n19; by Grainger, 62, 83; by Pennant, 38; by Thomson, 8, 29, 35, 37–38, 42; by Virgil, 48
Georgics (Virgil), 48
Gibson, Susannah, 64–65
Gigante, Denise, 114
Gillray, James, 114
Godwin, Catherine Grace, 199–200
Godwin, William, 90
Goldsmith, Oliver, 52, 97
Golinski, Jan, 47

Gosse, Philip Henry, 152
Grainger, James, 62–63, 83, 226n39
Gray, Samuel Frederick, 78
Gray, Thomas, 49, 51
Greek mythology, 81
Green, Thomas, 99
grief, 136
Groom, Nick, 117

Haitian Revolution, 88, 101
Hales, Stephen, 40, 65
Hamilton, William Rowan, 196–97, 199
Hartman, Geoffrey, 176
Harvey, William Henry, 152
Haughton, Hugh, 177
Hayley, William, 53, 217n57
Hazlitt, William, 4, 6, 19, 58, 174, 189, 190–92, 196, 205
Hemans, Felicia, 30, 64, 79–83, 199, 223n72
Hill, John, 178
Histoire Naturelle (Buffon), 102
History of Ireland (Moryson), 113
Hobbes, Thomas, 35, 46
Home, Henry, 19
"Horologe of the Fields, The" (Smith), 72–78, 79
Horologium Florae. *See* floral clock theory
Hortus Siccus, 182
Howard, Luke, 174
humor, 29, 35, 37, 53, 59, 94, 139–40, 144
hybridity, 87–88, 89, 91, 94, 125–28, 214n112, 226n51, 230n52, 231n58

Iceland, 42, 44
insanity. *See* madness
"Inscription Written on an Hermitage in One of the Islands of the West Indies" (Riddell), 108–9
Introduction to Botany (Lee), 25–26, 178
Introduction to Botany, An (Wakefield), 26
Ireland: Act of Union with Britain, 30–31, 110, 112, 117, 118; Irish Rebellion (1798), 114–15; potatoes and peasantry of, 116–20, 230n39

irritability of plants, 221n39
"I Wandered Lonely as a Cloud" (Wordsworth), 173–76

Jackson, Noel, 19, 174, 202
jalap, 63
Jewsbury, Maria Jane, 199
"Joan of Arc" (De Quincey), 204
Johnson, Claudia L., 11
Johnson, Shelby, 96
Jussieu, Antoine-Laurent de, 27, 187, 223n69

"Kalendar of Flora, The" (Smith), 137
Kames, Lord, 19
Kaul, Suvir, 38
Keats, John, 161, 176
Keith, Jennifer, 11, 20
Kelley, Theresa M., 25
Kelly, Gary, 79
Kennedy, Deborah, 199
Kiberd, Declan, 110
King-Hele, Desmond, 107, 160, 168, 174, 219n85
Kirkpatrick, Kathryn, 125
Kitson, Peter J., 197
Koerner, Lisbet, 64
Krueger, Misty, 96

Labbe, Jacqueline, 13
Lamb, Charles, 156
Landesborough, David, 152
law of dissociation, 233n47
Lay of an Irish Harp, The (Owenson), 121
Lectures on the English Poets (Hazlitt), 189, 190–91
Lee, James, 25–26, 178
Lee, Nathaniel, 173
Lessons for Children (Barbauld), 50
Letters on the Elements of Botany (Rousseau), 26, 122
"Letters to a Young Man Whose Education Has Been Neglected" (De Quincey), 192, 196–97, 206
Leviathan (Hobbes), 35

Lew, Joseph, 129
Lichfield Botanical Society, 26, 29, 35, 53, 170
Life of Erasmus Darwin (Darwin), 14
Lightfoot, John, 75
Lindley, John, 27, 223n67
"Lines Written in Early Spring" (Wordsworth), 20
Linnaeus, Carl: botanical classification system by, 4, 25–26, 32, 55, 138, 187, 225n29; Darwin and, 4; "The Delights of Nature," 55, 218n67; floral clock theory of, 29–30, 60, 63, 221n28; *Philosophia Botanica*, 62, 81; Romantic-era literature and botanical methods of, 24–27; *Systema naturae*, 25; *Systema vegetabilium*, 26, 29, 35; on transmutationist botany, 220n28
Linnaeus, Elizabeth, 174–75
Linnean Society, 26, 27, 70, 223n67
Literary Women of England, The (Williams), 79
"literature of knowledge" vs. "literature of power," 4, 9, 17, 19, 185, 189, 190, 192, 195–96, 202–7
Lloyd, Charles, 160, 161
Lobb, Richard, 121
Locke, John, 24, 77
Logan, Maria, 168
London Magazine (publication), 179
Looser, Devoney, 208
Loves of the Plants, The (Darwin), 4, 10, 15, 16, 64, 115. See also *Botanic Garden, The* (Darwin)
Loves of the Triangles (Canning), 159
Lunar Society of Birmingham, 15, 220n20
Lyrical Ballads (Wordsworth), 1, 4, 5, 14, 16, 17, 22, 65, 175, 176, 201

Mad Kate (Fuseli), 170, 171
madness, 155, 157, 167–68, 173–76, 193–94, 237n68. See also sanity
Magnolia grandiflora, 17
Mahood, M. M., 175, 179

Manchester Literary and Philosophical
    Society, 219n13
Mandeville, Bernard, 35, 46
maniock root, 101–2
marginality, 137–38
Martyn, Thomas, 26
masculinity, 8–9, 13. *See also* gender
Mathias, Thomas James, 16, 26, 194
May (goddess), 48, 55
McGann, Jerome, 3
McNeil, Maureen, 66
mechanical descriptions of plants, 29, 65,
    70. *See also* floral clock theory
mechanical flowers, 61
mechanism, 61, 65–72, 77, 78, 80, 220n18,
    221n35, 221n39, 223n63
Mellor, Anne, 3, 13
Memoir of John Aikin (Aikin), 27
Memoirs of the Life of Dr. Darwin
    (Seward), 31, 156, 157, 159
mental illness, 181–82, 184
Merian, Maria Sibylla, 97, 98
Metrical Miscellany (Riddell), 30, 88, 106–9
Milne, Colin, 75
Milton, John, 161, 195, 205
Mimosa pudica, 61, 62
Minor Morals (Smith), 137
Miss Hibernia at John Bulls Family Dinner
    (Cruikshank), 117, *118*
Mitford, Mary Russell, 200
Modest Proposal, A (Swift), 113
Montagu, Mary Wortley, 95, 123
Montesquieu, Charles-Louis de Secondat,
    baron de, 36, 57
morality, 36, 122, 161, 162, 190, 211n46,
    216n33. *See also* progress vs. decline, in
    British discourse
More, Hannah, 12, 192
"Morning Walk, The" (Clare), 58, 178
Moryson, Fynes, 113
Mullan, John, 11
Mundy, Francis Nöel Clarke, 54–55,
    217n62, 218n64
Myrone, Martin, 170
mythology, 48, 49, 55, 81

Natural History of Birds (Smith), 153
Natural History of Selborne (White), 119
naturalist's gaze, 123–25. *See also* poet-
    naturalists
Needwood Forest (Mundy), 54, 218n64
nerve theory, 19, 21, 24, 31, 156, 160–67
*New Illustration of the Sexual System of
    Carolus von Linnaeus* (Thornton), 51, 71
Newton, Isaac, 37, 204, 205, 219n83
Nie, Michael de, 115
"Nightmare, The" (Fuseli), 170
North, Julian, 198
novelty, 9, 10, 17–18, 203
Nugent, Maria, 100

"Ode on the Backwardness of Spring"
    (Penn), 50–51, 52
"Ode on the First of April" (Warton),
    216n39
"Ode to Mr. Gray" (West), 48
"Ode to the Olive Tree" (Smith), 151
"Ode to the River Darwent" (Darwin),
    106, 107–8
"Of Great and Little Things" (Hazlitt), 192
"On the Backwardness of Spring" (Bar-
    bauld), 47, 48–50, 217n44
"On the Life and Writings of Erasmus
    Darwin" (Cary), 179
opium, 167–68
originality: Aikin on, 7–8; of Darwin,
    15–16, 212n63; in poetry, 4–10; in
    scientific literature, 2, 9, 205–6; Seward
    on, 16, 21; in Smith's "Flora," 143–46. *See
    also* plagiarism
Ornithogalum umbellatum, 75
Ottum, Lisa, 147–48, 174
Ovid, 115, 138
Owenson, Sydney, 109; on colonial can-
    nibalism, 112–15; *The Lay of an Irish
    Harp*, 121. *See also Wild Irish Girl, The*
    (Owenson)

Packham, Catherine, 21, 65
Pancratium maritimum, 183
Papaver somniferum, 167

Pascoe, Judith, 142
Payne, Frances, 89
Penn, John, 47, 50–53
Pennant, Thomas, 10, 38
Percival, Thomas, 65, 70, 158, 166, 219n13
personification, 20
Peter the Great, 44
Petit Souper a la Parisienne, Un (Gillray), 114
Petley, Christer, 100
Philosophia Botanica (Linnaeus), 62, 81
Philosophy of Natural History, The (Smellie), 89, 91, 140–41
phytodynamism, 137, 229n24
Phytologia (Darwin), 70, 74, 81, 115, 166
Pinch, Adela, 176
"Pinna" (Smith), 150–51
plagiarism, 7–8, 13, 145, 159, 166, 205. *See also* originality
plant migration, 137, 229n24
plant studies. *See* botanical classification system; botanical writings; floral clock theory
Pleasures of Melancholy, The (Warton), 216n41
Pliny the Elder, 67
Poems (Owenson), 113
Poems (Wordsworth; 1807), 23
"Poetical Courtship, A" (Darwin), 107
poet-naturalists, 7–8, 26, 31. *See also* botanical writings; Darwin, Erasmus; naturalist's gaze; Smith, Charlotte
"Poet's Epitaph, A" (Wordsworth), 18–19
Pole, Elizabeth, 107
Political Justice (Godwin), 90
Polwhele, Richard, 26, 120, 171
Pomona (goddess), 48
Pope, Alexander, 139, 140, 142
poppy, 167–69, 235n30
Porter, Dahlia, 28
Porter, Roy, 155–56
potatoes, 116–20
Pratt, Mary Louise, 124
primitivism, 104, 215n24
Principia (Newton), 204, 205

progress vs. decline, in British discourse, 104, 215n24, 216n31
Pursuits of Literature (Mathias), 16, 26, 194

Questioning Nature (Bailes), 2, 5, 7, 187, 203

racism. *See* scientific racism
Rape of the Lock, The (Pope), 139, 140
Rawdon, Elizabeth, 113
"Reflections on Some Drawings of Plants" (Smith), 136
Reflections on the Revolution in France (Burke), 114
Reill, Peter Hans, 65, 66
Reno, Seth T., 174
Richardson, Alan, 12
Riddell, Maria, 30; biography on Burns by, 105; on cashew tree, 99–100, 226n39; Darwin and, 105–6; "Elegy on the Death of Captain J. Woodley," 108; "Inscription Written on an Hermitage in One of the Islands of the West Indies," 108–9; *Metrical Miscellany*, 83, 105–6; slavery and, 101, 102, 104–5; on Smellie, 92–93; *Voyages*, 30, 87, 88, 89, 94–95, 106
Riskin, Jessica, 24
Robinson, Daniel, 145
Robinson, Henry Crabb, 200
Rodgers, Nini, 116
Roman mythology, 48, 49, 55
Romanticism: Linnaean botany and, 24–27; sensibility and, 11–14, 21, 211n46; Wordsworth's preface on Darwin and, 14–24. *See also* aesthetics of Romantic-era literature
Rose, Hugh, 73
Ross, Marlon, 3, 209n14
Rothstein, Eric, 19–20
Rousseau, G. S., 12
Rousseau, Jean-Jacques, 26, 122, 173
Royal Society, 26
Rural Muse, The (Clare), 58
Russia, 42
"Ruth" (Wordsworth), 17

Sabine, Marie, 102
Sacred Theory of the Earth (Burnet), 59
Salaman, Redcliffe, 117
sanity, 155–57. *See also* madness
"Sanity of True Genius, The" (Lamb), 156
Saville, John, 157
Schaw, Janet, 87, 94, 100
Schiebinger, Londa, 25, 98, 212n63
scientific literature: defined, 2; gender and, 8–9, 23, 25, 26; Romantic-era sentiments on, 1–2, 22–23, 187–208; sensibility and, 2–3; women and, 8–9, 23, 25, 26, 195–97. *See also* botanical writings; Darwin, Erasmus
scientific racism, 2, 32, 102–4, 188–89, 205, 227n63
scientific sexism, 2, 32, 188–89, 205
Scott, John, 5
Scott, Walter, 173
Scottish Enlightenment, 45
seasonal temporality, 29. *See also* temporal backwardness; temporal sensibilities
Seasons, The (Thomson), 8, 29, 35–44, 58–60
sensation, 19, 21, 24, 31, 156, 160–67
sensibility: Chandler on, 211n46; Darwin and, 19, 21, 25; Owenson on, 121–22; in poetry, 5–7; Romanticism and, 11–14, 21; scientific literature and, 2–3; Seward on, 31–32, 156–57; Thomson on, 35–44, 215n17, 215n19; Wilson on, 235n19. *See also* truth
Sensitive-Plant (Shelley), 61, 222n61
sensitive plant studies, 24–25
sentiency of plants, 70, 71, 115
sentiment, 10, 12, 106, 121–22, 162–63
sentimental love, 36, 79, 163, 237n68
Seward, Anna, 5; on Darwin's *Botanic Garden*, 10; *Elegy on Captain Cook*, 157; illness of, 181–82; literary criticism by, 156, 157–60; *Memoirs of the Life of Dr. Darwin*, 31, 156, 157, 159; on originality, 16, 21; on sensibility, 31–32, 156–57; "To the Poppy," 168–69, 170; on Wordsworth's poetic insanity, 173–76. *See also*

"Backwardness of Spring Accounted For, 1772, The" (anonymous)
sexism. *See* scientific sexism
sexual system of plants, 25–26, 54, 74, 90–94, 120, 133, 141–42, 178, 221n35. *See also* botanical classification system
Shakespeare, William, 108, 161, 180, 195, 205, 234n2
Sharpe, Charles Kirkpatrick, 89
Sharpe, William, 156
Shelley, Percy, 61, 62, 219n3, 222n61
Shepherd's Calendar, The (Clare), 99
Shteir, Ann B., 25, 26
slavery, 100–105, 116
Smellie, William, 89, 91–92, 101, 140–41
Smith, Charlotte, 5; on agency of plants, 29; "Beachy Head," 142, 152, 222n54; *Beachy Head, Fables, and Other Poems*, 72, 77; on botanical clock, 64, 71–72, 74–75; *Conversations Introducing Poetry*, 31, 133, 135; Darwin and, 137–39; *Elegiac Sonnets*, 74, 136, 138; "Flora," 31, 72, 133–43, 146, 147, 152–53; "The Horologe of the Fields," 72–78, 79; "The Kalendar of Flora," 137; *Natural History of Birds*, 153; "Ode to the Olive Tree," 151; "Pinna," 150–51; on poetic achievement, 21–22; "Statice," 147; "St. Monica," 13; "Studies by the Sea," 133, 137; writing style of, 12–13
Smith, James E., 27, 78
"Snowdrop, The" (Thornton), 51
Social History and Influence of the Potato, The (Salaman), 117
Solander, Daniel, 44
"Sonnet on the Death of Richard West" (Gray), 49
Specimens of British Poetesses (Dyce), 200
"Speculations on the Perceptive power of Vegetables" (Percival), 65
Spenser, Edmund, 113
Spinoza, Baruch, 174
Spring. See Seasons, The (Thomson)
stadial theory, 36, 56
Stafford, Barbara, 148

INDEX

"Stanzas . . . Off St. Bees' Head" (Wordsworth), 13
starvation, 113, 116–20
"Statice" (Smith), 147
Stillingfleet, Benjamin, 50
St. Kitts, 89
"St. Monica" (Smith), 13
"straw man," as concept, 1–2, 22
"Studies by the Sea" (Smith), 133, 137
Sugar-Cane, The (Grainger), 62, 226n39
Summer. See *Seasons, The* (Thomson)
Sweet, John Wood, 102
Swift, Jonathan, 113
sylphs, 140, 232n19
Systema naturae (Linnaeus), 25
Systema vegetabilium (Linnaeus), 26, 29, 35

Task, The (Cowper), 135, 169–72
Taylor, Jesse Oak, 42
Temple of Nature, The (Darwin), 17, 56, 69
temporal backwardness, 44–47. See also backwardness of spring
temporal sensibilities, 35–44, 215n17, 215n19, 219n83. See also sensibility
Thomson, James: about, 5, 39, 225n27; Penn on, 52; *The Seasons*, 8, 29, 35–44, 58–60, 219n83; on temporal sensibilities, 35–44
Thornton, Robert, 155
Tickell, Thomas, 201
"To a Geranium Which Flowered during Winter" (Smith), 137
"To Opium" (Logan), 168
torrid zone, 29, 36, 37, 41–42, 216n26, 216n33. See also climate and sensibility
"To the Poppy" (Seward), 168–69, 170
Townson, Robert, 70
Tracy, Thomas, 126
trans, as prefix, 87
transmutationist botany, 220n28
Travels (Bartram), 17–18
travel writings. See Riddell, Maria
truth: Aikin on, 8, 142, 213n91; De Quincey on, 203; poetry and, 191; Wordsworth on, 17, 22. See also sensibility

tulipomania, 155
Tupper, James Perchard, 78
Turkish Embassy Letters (Montagu), 95, 123

Uberti, Fazio Dagli, 228n8, 229n22
United Kingdom: Act of Union with Ireland, 30–31, 110, 112, 117, 118; colonialism of, 30–31, 87–88, 94–105, 128–30; political cycles in, 214n4; on progress vs. decline, 104, 215n24, 216n31; Thomson on temporal sensibilities of, 35–37, 39
Universal Herbal (Green), 99
Unsex'd Females (Polwhele), 26, 171

Van Sant, Ann Jessie, 12
Vegetable Staticks (Hales), 40, 65
Venus (goddess), 55, 139
"Verses Supposed to Have Been Written in the New Forest, in Early Spring" (Smith), 136–37
View of the Present State of Ireland, A (Spenser), 113
Virgil, 48, 51
vitalist theory, 70
Voltaire, 115
Voyages to the Madeira, and Leeward Caribbean Isles (Riddell), 30, 87, 88, 89, 94–95, 106

Wakefield, Priscilla, 26
"Walk in the Shrubbery, A" (Smith), 136
Walpole, Horace, 90, 171
Warton, Thomas, 216n39, 216n41
Watts Academy, 37
Wedgwood, Tom, 168
West, Richard, 47, 48
West Indies, 87–88, 94–105
Whale, John, 204
Wheeler, Roxann, 103
White, Gilbert, 119
Whitehead, James, 155, 180, 184
Whitehead, William, 5

Wild Irish Girl, The (Owenson), 30–31, 110–12; colonialism and class consciousness in, 128–30; hybridity in, 125–28; naturalist's gaze in, 123–25; potatoes and peasantry in, 116–20; sensibility in, 121–22; sexuality and plants in, 120–21, 122–23
Williams, Helen Maria, 5, 158
Williams, Jane, 79
Willis, Thomas, 24
Wilson, Kathleen, 103
Winter. See *Seasons, The* (Thomson)
"Winter" (Kent and Tardieu), 43
Withering, William, 26, 75–76, 127, 140
Wollstonecraft, Mary, 12, 89, 171, 224n8
woodland, poetic portrayal of, 143–46
Woodley, William, 89
Wordsworth, Dorothy, 175
Wordsworth, William: "A Poet's Epitaph," 18–19; definition of poetry by, 14; egotism of, 154, 156, 172–76, 191, 236n54; "Essay, Supplementary to the Preface," 202–3; on gender and science, 23–24; "I Wandered Lonely as a Cloud," 173–76; "Lines Written in Early Spring," 20; on literature and science, 22–23, 191–208; *Lyrical Ballads*, 1, 4, 5, 14, 16, 17, 22, 65, 175, 176, 201; on personification, 20–21; on philosophy, 223n63; *Poems* (1807), 23; preface on Darwin and Romantic-era literature, 14–24; "Ruth," 17; sexism by, 3, 23–24, 199; Smith's writing style and, 12–13; targeting Darwin by, 1, 15–23, 201
"Works of Alexander Pope, The" (De Quincey), 202

yellow fever, 101, 144
Young, Edward, 156

Zoonomia (Darwin), 16, 31, 98, 156, 162–65, 194, 224n15

CPSIA information can be obtained
at www.ICGtesting.com
Printed in the USA
LVHW040759070723
751646LV00003B/246